electronics drafting

by
John Frostad
Professor of Technology
Green River Community College
Auburn, Washington

South Holland, Illinois
THE GOODHEART-WILLCOX COMPANY, INC.
Publishers

Library of Congress Catalog Card Number 91-22600
International Standard Book Number 0-87006-911-X

3 4 5 6 7 8 9 10   92   97 96 95 94

Library of Congress Cataloging in Publication Data

Frostad, John.
  Electronics drafting / by John Frostad.
    p. cm.
  Includes index.
  ISBN 0-87006-911-X
  1. Electronic drafting. I. Title
TK7866.F76 1991
621.381--dc20   91-22600
                                                  CIP

INTRODUCTION

ELECTRONICS DRAFTING provides you with basic instructions in electronics drawing. It begins with a review of drafting fundamentals including drafting tools, inking methods, and taping methods. Electronics drafting topics include the meanings of electronics terms, electrical component descriptions, electronic component symbols, reference designations, and PC board conductor spacing. You will learn the proper use for symbols in block diagrams, flow diagrams, single line diagrams, schematic diagrams, and logic diagrams.

ELECTRONICS DRAFTING teaches you to think, create, and draw in a logical sequence. You are asked to begin with a sketch. From this, you will generate all of the formal schematic drawings, parts lists, wiring designation lists, printed circuit board layouts, PC board artworks for photoresist exposure, PC board marking artworks, and packaging drawings.

ELECTRONICS DRAFTING is written in concise, easy-to-read language and illustrated with hundreds of photos and drawings to help you learn. The book typically provides you with two visual examples for each major topic covered. You will have the opportunity to complete many problems to aid in your learning experience. The questions provided at the end of each chapter emphasize the important information covered in the text.

John R. Frostad

USING THIS BOOK

1. *Read the objectives* for the chapter. They will tell you what will be learned by studying the chapter.
2. *Preview the chapter.* Scan through the chapter headings. Read the illustration captions and look at the illustrations briefly.
3. *Study the chapter* carefully. Refer to and examine the illustrations as soon as they are called out.
4. *Complete the review questions.* Try to answer as many as you can without referring back into the chapter. Restudy any that you cannot answer from memory.
5. *Work the electronic drafting problems.* Ask your instructor for any additional information about doing the problems.
   A. Obtain the *correct materials* for the problem: paper or vellum, pencil or pens, drafting instruments, etc.
   B. If needed, make a *rough sketch* of the project in pencil. Your instructor may want to check your work before you progress too far.
   C. With some problems, you may need to *prepare outlines, starting points,* or *draw light ruled lines.* Sometimes, this may be done before coming to class.
   D. Make sure you *know all of the details* of the problem before doing too much work. Your instructor may hand out information sheets for each problem or may verbally give details.
   E. *Ask your instructor* for help with a problem if you are in doubt.

CONTENTS

Chapter 1

INTRODUCTION TO ELECTRONICS DRAFTING

After studying this chapter, you will be able to:
- List the responsibilities of the engineer.
- State what a drafter does in electronics.
- Describe how form follows a part's function.
- List the job levels in drafting.

In order to understand the drafter's responsibilities, you must first study how an engineering organization is set up. Fig. 1-1 shows such an organization of a typical company. Everyone in the organization has responsibility for specific tasks. You must understand the responsibilities of other people in order to work well with them as a team member.

ENGINEER'S RESPONSIBILITIES

The engineer may work from an idea he/she originates. But in most cases, the engineer gets directions from engineering, service, manufacturing, marketing, and purchasing managers.

Marketing, for example, will submit requests for new products or the redesign of existing products. Service will inform engineering about maintenance problems. They will record all service failures and suggest servicing requirements. Manufacturing will request engineering support to minimize manufacturing costs. They may need assembly tools, or some parts redesigned to make them less costly. The Electronics Engineering department is asked to design products for a highly competitive market. Designing better products and making them less expensive to build is the responsibility of all departments. Engineers must also work with the drafting department. They are responsible for providing the drafter with most of the information necessary to create a drawing. Engineers will submit sketches, layouts, and other written instructions. The amount of instructions given to a drafter will normally depend on their knowledge and background. Engineers will not give a drafter a task that cannot be completed.

DRAFTER'S RESPONSIBILITIES

Drafters are responsible for preparing drawings that are easy to read. The drawing should completely describe the engineer's design requirements. It

PRESIDENT				
MARKETING MANAGER	ENGINEERING MANAGER	MANUFACTURING MANAGER	PURCHASING MANAGER	SERVICE MANAGER
Provide marketing forecasts. Secure contracts for product development. Investigates competitor's products. Advertises and sells product. Directs engineering in product design.	Designs products. Researches new products. Supports marketing, manufacturing, purchasing, and service. Prepares drawings and maintains product and design history. Studies product reliability.	Produces the product. Creates cost reduction for products. Inspects and tests products. Produces jigs and fixtures. Schedules production sequence. Consults engineering on manufacturing requirements.	Purchases parts, equipment, and raw material. Schedules materials, parts, and equipment to enter the assembly at set time. Recommends materials and parts for engineering and manufacturing.	Analyzes service problems. Recommends product redesign. Service products. Writes service and operations manuals. Consults engineering on service requirements.

Fig. 1-1. A typical engineering organization would have the listed departments. Each department is responsible for the duties listed. Note: All departments work with each other to benefit the organization.

must also meet company and customer drawing requirements. Drawings will allow the reader only one interpretation. You cannot leave decisions to the reader.

Examine Figs. 1-2 through 1-5. They will show some engineering inputs and the drafter's finished design work. In designing electronics equipment, an engineer and drafter are concerned with the:

1. Part functioning correctly.
2. Safety of the operator and equipment.
3. Equipment being a pleasing design to the customer.

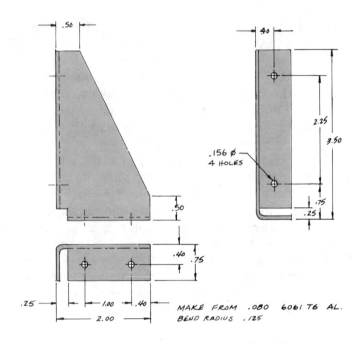

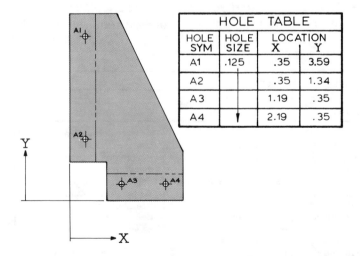

Fig. 1-2. The engineer's input to the drafter.

HOLE TABLE			
HOLE SYM	HOLE SIZE	LOCATION	
		X	Y
A1	.125	.35	3.59
A2		.35	1.34
A3		1.19	.35
A4		2.19	.35

Fig. 1-3. The drafter's finished drawing from Fig. 1-2. The drafter developed a flat pattern with a hole table to control the engineer's part. This drawing will also meet the manufacturer's document requirement.

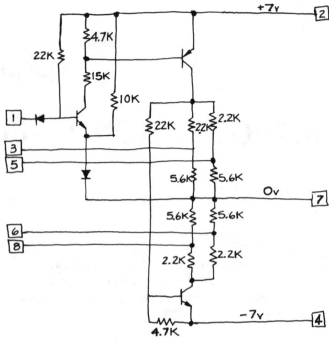

NOTE:
UNLESS OTHERWISE STATED
RESISTORS ARE RATED IN OHMS & 1/4 W
PNP TRANSISTORS ARE 2N3644
NPN TRANSISTORS ARE 2N3643
DIODES ARE 1N662A

Fig. 1-4. This is an example of the engineer's input to the drafter. From this, a schematic drawing can be generated.

4. Equipment's environmental needs, which may mean a special electrical supply, cooling, or dust-free room.

These suggest some of the decisions a designer must make.

The drafter must consider the housing for a system. Electronics equipment will normally be housed in plastic or metal packages. On the outside of the package (enclosure) will be switches, knobs, lights, meters, and other components. On the inside, circuit boards, electrical and electronic components, wiring, and other devices must be appropriately positioned and secured. The shape of the package will be dictated by the design function or use.

Calculators are a good example of how form or shape follows function. Fig. 1-6 shows a small solar calculator. It is small enough to carry in your pocket and it performs all the basic math functions. Many people carry a calculator of this sort with them when they go shopping. Fig. 1-7 shows a larger and more powerful calculator. This is the calculator most often used by engineers and drafters. Note: the keyboard matches the size of the fingers.

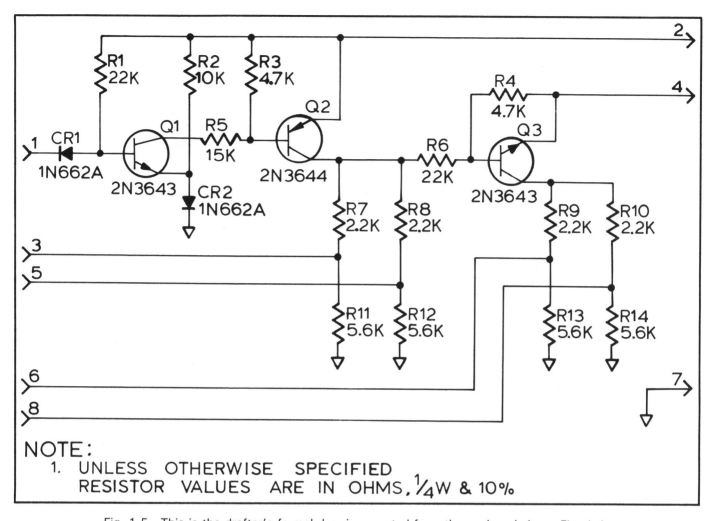

Fig. 1-5. This is the drafter's formal drawing created from the engineer's input Fig. 1-4.
Note: The drafter is responsible for the correctness of the symbols and the overall presentation.

Fig. 1-6. A small calculator designed for solving simple arithmetic problems.

Fig. 1-7. A calculator with a full array of functions is the type used by most drafters and engineers.

The desk calculator shown in Fig. 1-8 is used in business. It has the inconvenience of being harder

to transport. All other features are developed for the user's convenience. It has large keys, which are appropriately spaced for easy manipulation. It provides a large display area for easy reading of the input and

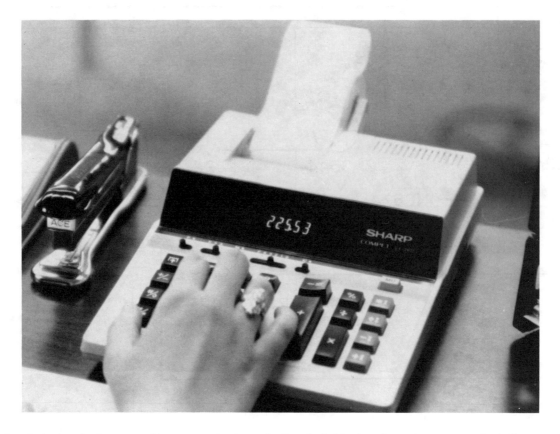

Fig. 1-8. A desk calculator used by accountants and other individuals who enter many calculations each day.

output information. It has a printout so the entries can be checked and a record kept of the calculations. The unit will be heavy enough to sit firmly on a desk while being used.

An accountant or business person who uses a calculator many hours each day would not be without these features. The decision on what type calculator to manufacture is a marketing task. Marketing will take a survey of what the customers are buying or want and then will suggest to engineering what to design. Marketing predictions influence much of the designing decisions made by engineers and design drafters.

JOB LEVELS IN DRAFTING

Education is the first step in the career of a drafter. Electonic firms are able to use drafters from all levels of drafting programs. High schools, technical schools, community colleges, and four-year technical programs all train drafters for industry. An education period is required by many companies. They want an assurance that they are hiring a drafter with a solid background. This training background can be demonstrated by a well compiled portfolio of your work. The portfolio should include different types of drawings and technical

reports you have done. It is meant to show your skills, demonstrate your knowledge, and give a future employer an idea of your educational background. The employer will establish the level of work you can do by studying your portfolio. In addition to your portfolio you should prepare a RESUME (re-zu-may').

Drafters from beginners to veterans will use a resume in their job search. It will normally be required before an interview can be arranged. What should you put in a resume? It should describe work history, educational background, special interests, accomplishments, and personal data. Never tell the employer all about yourself in your resume. Touch on the highlights just enough to generate interest in you. This interest will get you an application form or personal interview. You will then use this application or interview to tell about yourself. A catchy resume has helped many drafters get a job. Remember to be brief, unique, and you may be the drafter getting an interview.

DETAIL DRAFTER

The detail drafter's position is where most drafters begin. The position is sometimes referred to as detailer. The detailer will work under close supervision. This supervision will continue as long as the detailer is learning the fundamentals of drafting. While learning these fundamentals, detailers will be doing drawing revisions, detail drawings, parts lists, schematics wire lists, simple assemblies, and other knowledge-building tasks.

ASSEMBLY DRAFTER

The assembly drafter is responsible for the fit and form of each part of the equipment. They will draft the assembly drawing which shows how all the individual parts come together. When fitting these parts into the assembly, they will check for part interferences, mounting hole alignment and the compatibility of mating parts. The assembly drafter will work under the supervision of an engineer and designer. In this position, you will learn:
1. Additional drafting skills.
2. How to use manufacturer's catalogs.
3. About machining and machine shop practices.
4. How to dimension and tolerance parts.
5. Additional engineering mathematics.
6. How to work with engineers.
7. How to work independently.

When an assembly drafter demonstrates an understanding of all these tasks, he/she will be ready to become a designer.

DESIGNER

The designer position is the highest level in drafting. Some companies have different levels of designers. In those companies a senior designer is the top position. Designers are required to know about shop practices, engineering mathematics, quality control of manufactured calculations, and other engineering functions as required by the different companies. Designers will work with and supervise other drafters. They work as members of the engineering design team. They consult with engineers and incorporate their ideas into the drawing.

CHECKER

All companies require that each drawing be checked when completed. Checking can be formal or informal. Informal checking may simply require that drafters review their own work. Other companies will have drafters check each other. But formal checking will be done by an experienced drafter. Their checking will involve:
1. An inspection of linework and lettering quality.
2. A check to insure that the drawing meets all company drafting standards.
3. A study to see that all parts meet form, fit, and

function requirements.
4. A check to determine manufacturing feasibility—can the part be economically manufactured?

Checkers are drafters with years of experience. They must understand engineering standards and manufacturing methods. They are normaly equal to the designer in status. Checkers can suggest design changes when they see a change is required. If no changes are required and all company standards are met, the checker will sign his/her name in the drawing's title block. The checker's signature implies that the drawing is ready to be sent to production, inspection, and other drawing users.

Drafters are required to submit their drawings to the checker along with the following information:
1. All written information supplied by the engineer.
2. All preliminary layouts.
3. All mechanical calculations.
4. Reference prints.
5. Any other necessary data.

The checker cannot adequately check your drawing without all the facts.

ILLUSTRATOR

Most electronics drafters can do technical illustration drawings. Fig. 1-9 shows the information given to the drafter. Fig. 1-10 shows the finished drawing. These drawings are frequently used for assembly drawings and for customer service and operating manuals. Some companies have a "manuals" department where the drafters specialize as illustrators. They will draw isometric drawings, graphs, block diagrams, and charts. They create graphic presentation work such as advertising or patent drawings.

DRAFTING SUPERVISOR

Drafting departments are normally assigned a supervisor. This position requires an excellent knowledge of drafting and years of experience. The supervisor has the responsibility for scheduling drafting work, calculating personnel needs, maintaining and originating drafting standards, and creating or continuing a good working relationship between drafting and engineering. Supervisors must be able to evaluate and make regular performance reports on each drafter. They must also be good leaders so they can help motivate drafters to reach their highest potential.

TYPES OF COMPANIES

Electronic companies may be divided into different types. See Fig. 1-11. One group is the FUNCTIONALIZED or PROJECTIZED company. Another

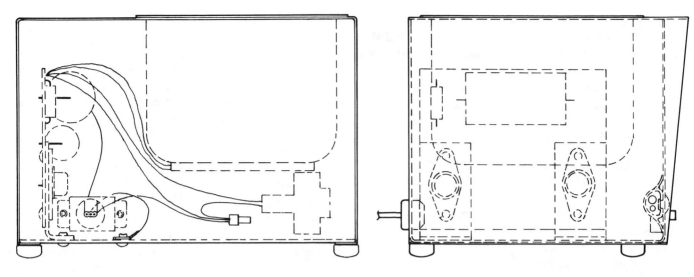

Fig. 1-9. A typical input to an illustrator. The illustrator will often have to research the hardware in order to finish the drawing. See Fig. 1-10 for the illustrator's drawing.

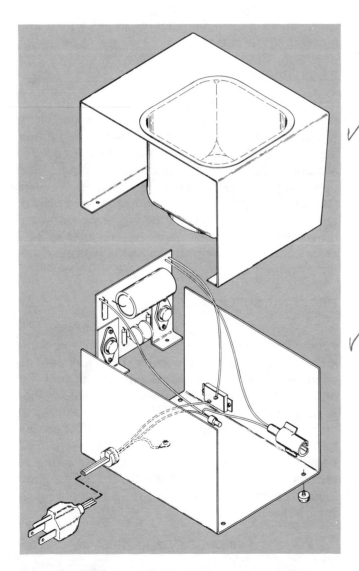

Fig. 1-10. The illustrator's drawing of Fig. 1-9. Note: It was decided by the illustrator to make an exploded isometric drawing. This is often the type drawn.

includes government contractors or nongovernment contractors. Knowing about the companies and the way they work is important. The type of work and work assignments can change the level of interest you have in your job.

FUNCTIONALIZED COMPANIES

A functionalized company will most often be a larger firm. These firms have enough specialized work to hire drafters to do specific tasks. An example could be a drafter hired to draw only schematics. The work in other areas will be accomplished by other specialized drafters. Functional-oriented companies can normally use drafters with less training because it takes fewer hours to train a drafter to perform a single task.

PROJECTIZED COMPANIES

Most smaller companies assign their drafters to projects. A drafter working on a project will do most of the drawings required. Drafters will coordinate their drawing requirements with the project engineer, lead drafter, technicians, and others working on the same or similar projects.

Project companies prefer to hire drafters with a broad training background. This will enable the drafter to work on the many job tasks. Drafters with project experience find it easier to move from job to job. Their broadened background and work experience will help them secure employment. Refer again to Fig. 1-11, which suggests how the company can help a drafter, designer, or engineer grow in understanding and background.

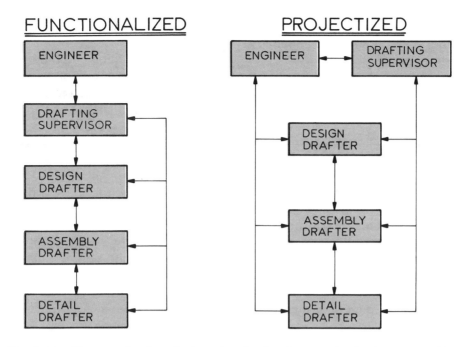

Fig. 1-11. The organization of a functionalized and or a projectized company. Note: In most functional organizations, the detail drafter will seldom communicate with the engineer; but in the projectized company, there is communication at all levels.

GOVERNMENT CONTRACTORS

Many companies design electronics equipment for the Defense Department or for the United States Government. Companies receiving government contracts must adhere to strict drafting procedures and standards. Fig. 1-12 shows some of the standards which control the drawings.

The government has standards written to control most projects for which they grant contracts. These standards help the contractor design and document electronics equipment.

NONGOVERNMENT CONTRACTORS

Companies who do not sell their drawings and equipment to the government usually are not so controlled in their documentation. They will, however, have company standards which are normally based on military standards. There are associations of government and nongovernment contractors which are constantly working on industry-wide standards. The purpose is to make interchangeable components and produce documentation familiar to all readers.

SKILLS REQUIRED

In most areas of industry there are requirements for training. Companies will have requirements for communication and technical skills.

COMMUNICATION SKILLS

The electronics drafter will exchange ideas with many people on the job. You will cooperate with manufacturing, service, marketing, engineering, and accounting personnel. You will not only communicate by drawings. Many times you will have to present information both written and verbal. Developing these skills in the classroom is important. It is important because they are the first skills

MIL-D-1000	Drawing
MIL-D-55110	Printed Wiring Boards
ANSI-Y-14.5M	Dimensioning and Tolerancing
USAS-Y14.15	Electrical and Electronic Diagrams
MIL-STD-12C	Abbreviations for use on Drawings
MIL-STD-429	Printed-Wiring Printed Circuit Terms
ANSI-Y32-2	Electrical and Electronic Diagrams, Graphic Symbols for
MIL-STD-275	Printed Wiring for Electronic Equipment
MIL-STD-429	Printed-Circuit Terms and Definitions
MIL-STD-454	Standard General Requirements for Electronic Equipment
USAS-Y32.14	Graphic Symbols for Logic Diagrams
USAS-Y32.16	Electrical and Electronic Reference Designations

Fig. 1-12. A list of documents that specify drafting standards for projects with the U.S. Government.

you use when you enter a company. The interviewer will judge you by your communications capability as well as your resume and portfolio.

TECHNICAL SKILLS

Many students entering electronics drafting classes believe they must fully understand electronics theory in order to work as an electronics drafter. This is not a requirement. Some knowledge of simple principles is required, but a full understanding is not. The engineer is responsible for doing the mathematics and designing of the circuit. They have many hours of education and training to enable them to do their task. Drafters are only required to create the drafting support and develop enough knowledge to produce electronics drawings.

DRAFTING EVALUATIONS

Drafters are normally evaluated on six basic factors: quality and quantity of work, knowledge of drafting, dependability, cooperation, and initiative. Your drafting knowledge and quality of your work will help you get a job. To keep a job you must be dependable, cooperative, show some initiative, and put out quality work. Initiative is the key word for job success. If you show some initiative by continuing your education, working toward professionalism, and cooperating with others, you will be treated well in your career.

REVIEW QUESTIONS

1. The electronic equipment should be designed so that the form follows _____.
2. How does marketing assist engineering?
✳3. What might engineers supply the drafter when they want a drawing created?
4. What is the normal starting position for a drafter?
5. List two things required by most companies when they interview a drafter.
6. What are four general things the checker looks for?
7. List the things an assembly drafter will learn.
8. Why are good communication skills required of the drafter?
9. List what the drafter should submit with the drawing when it is routed to the checker.
10. What six things are used to evaluate the drafter on the job?

STUDY ACTIVITY

1. Check the newspaper for drafting jobs. Select those for which you expect to qualify and state job descriptions, pay levels, and companies products if available.
2. Interview an experienced electronics drafter. Prepare a report on their background, training and job activities.
3. Call a company and ask for an application form. Fill out the application and turn it into your instructor to be checked.

The modern electronics drafter uses computer aided drafting design (CADD) equipment. The digitizing pen reads and transmits an object's location.   (Bendix Corp.)

Chapter 2

DRAFTING FUNDAMENTALS

After studying this chapter, you will be able to:
☐ List the standard drawing sizes.
☐ Explain the need for drawing quality.
☐ Identify some basic drafting tools.
☐ Explain the requirements for lettering.
☐ List the uses for the different line widths.
☐ Create an orthographic drawing.
☐ List basic parts of the computer hardware system.
☐ Define computer aided design (CAD).
☐ List some advantages of CAD.

The electronic drafting field requires high quality work. Drawings created by drafters will be used by others who will spend hours reading them. With the amount of time spent reading drawings, it makes sense to spend an extra few minutes making them easier to read. These readers may be your customers. Quality drawings are a good advertisement for a company's technical ability.

DRAWING STANDARDS

This chapter will cover the general drafting practices of industry. These practices will be based on the latest standards. Most companies closely follow military and industrial standards. The first standard you will study is that for lettering.

FREEHAND LETTERING

The name "freehand" does not mean without using templates. Some aids are used to set up horizontal guidelines.

The standard lettering style used in companies making electro-mechanical equipment is the single stroke upper case (capitals) gothic, Fig. 2-1. Most companies prefer this lettering to be vertical. The normal height of the letters will be 5/32 in. The reason for this style and height of lettering is to

Fig. 2-1. Typical upper case lettering style.

allow consistent microfilming. MICROFILMING is the process of photographically reducing the size of the drawings for documentation, storage, and fast retrieval. A microfilm of a drawing is retrieved and a "blow-back" made to get an original size drawing or blueline print.

Lower case lettering may be used for special applications. Some of the applications will be: connector pin identification, name plates, and switch positions. See Fig. 2-2 for single stroke lower case lettering.

Fig. 2-2. Typical lower case lettering style.

13

Some companies will permit you to use inclined lettering. If inclined lettering is used, it should have a uniform slant. The slant is normally 68 degrees, Fig. 2-3.

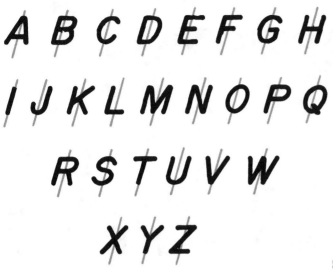

Fig. 2-3. Typical slanted lettering style.

Freehand lettering is used in most industries. You need to practice lettering until you develop good techniques. Get a feel for lettering composition. Composition is how the lettering looks in words and sentences. Use guidelines for freehand lettering, Fig. 2-4. One aid called an Ames lettering guide is popular. Guidelines should be drawn lightly so they are just visible at arms length.

MECHANICAL LETTERING AND TRANSFER LETTERING

When it is important to create a high quality drawing, mechanical lettering is sometimes used. One of the following three methods may be used:

1. Mechanical lettering is produced by using a scriber in a template, Fig. 2-5. With this tool, you can also create standard symbols and electronic symbols.
2. Typewriters may also be used to do lettering on drawings. You may also type on a transparent tape with an adhesive back. After the information has been typed, the tape can be cut apart and placed on the drawing.
3. Rub-on lettering comes on sheets in many styles. To put the lettering on the drawing, all you need is a dull pencil, ball-point pen, or similar tool. Position the letters or symbols on the drawing and rub over them to transfer them to the drawing, Fig. 2-6.

 ## LINEWORK

All lines that we wish to reproduce need to be dark (dense). Electronic drawings require linework to be dark and about .02 in. wide. To make dense lines of consistent width requires practice. Different line widths will help accentuate features on the drawing, Fig. 2-7.

Refer to Fig. 2-8. Inked drawings make it easy to get different widths of lines. You simply choose a different sized technical pen. Pencil drawings are more difficult. But with practice, in sharpening, rolling the lead holder as the line is drawn, and holding

A

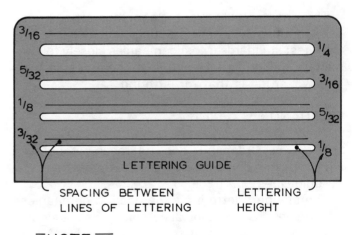

B

Fig. 2-4. A—Good freehand printing showing style and composition. B—Template for setting up lettering heights.

14

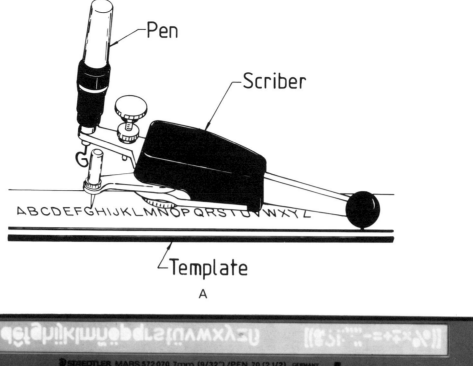

Fig. 2-5. A—Mechanical letters using a template and scribe. B—The template that guides the scribe.   (Staedtler Mars)

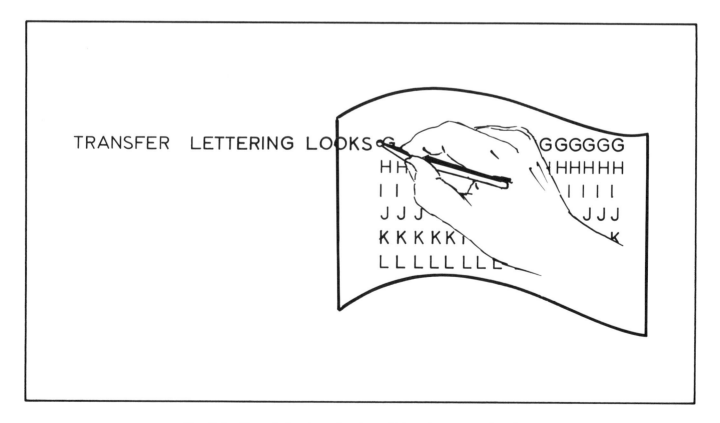

Fig. 2-6. Transfering lettering by rubbing them onto drawing.

LINE APPLICATION	LINE THICKNESS
GENERAL USE	MEDIUM .020-.025
MECHANICAL CONNECTION & SHIELDING	MEDIUM .020-.025
LEADER LINES	THIN .010-.015
BOUNDARY LINES OR GROUPING LINES	THIN .010-.015
LINES FOR EMPHASIS	THICK .030-.038

A

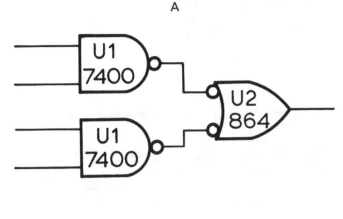

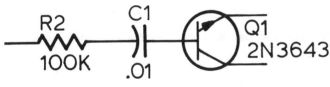

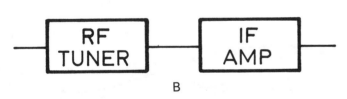

B

Fig. 2-7. A—Each type of line has a purpose in an electrical diagram. B—Often part symbols are emphasized with heavy line width.

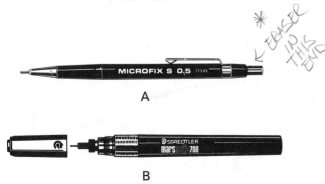

Fig. 2-8. These tools are used to create different and consistent line widths. A—Thin lead pencils. B—Technical pens. (Staedtler Mars)

it against the straight edge, you will learn to get quality pencil lines. Thin lead pencils are becoming more popular in industry. They make it easy to create consistent linework.

EQUIPMENT AND MEDIA

Electronics drafting requires many of the instruments used by drafters in other fields. The major difference is the group of templates used. All drafters need to know how to purchase, use, and maintain their drafting equipment. To purchase drafting tools, you may want to go to an engineering supply store that carries professional brands. When you purchase the tools, the salesperson will be able to explain the use and care of each item. Care will normally mean keeping the tools clean with mild soap and water.

TEMPLATES

As an electronics drafter you will draw the same symbol repeatedly. This repetition will require the use of templates, Fig. 2-9. Templates are designed to be used with either pencils or technical pens. Templates designed for pencil work will have wider guiding slots to allow for the wider pencil lead.

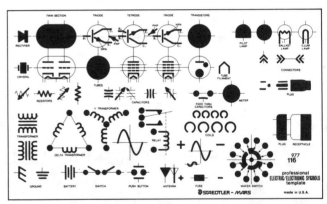

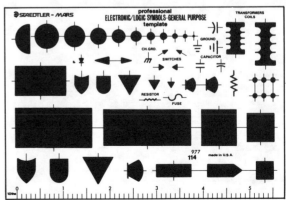

Fig. 2-9. Templates provide uniform symbols on electronics drawings. (Staedtler Mars)

Technical pens used in the pencil templates will create sloppy symbols. Both technical pens and thin lead pencils can be used in pen templates.

When purchasing a template, consider specification standards on how symbols are constructed. Good templates will meet military and industry requirements. Most companies purchase templates for their drafters to use. They do this so that each drafter will be able to work uniformly in their system.

TAPES

Tapes are used in many ways in electronics drafting. They can be applied to graphs, diagrams, schematics, and printed circuitry. The tapes come in many different widths, colors, and styles.

DRAFTING TABLES AND MACHINES

Engineering firms are concerned that drafters have a good environment in which to work. A good environment will increase the drafter's productivity. Most companies will supply all the drafter's needs except for some of the small items. See Fig. 2-10 for a view of a typical drafting station.

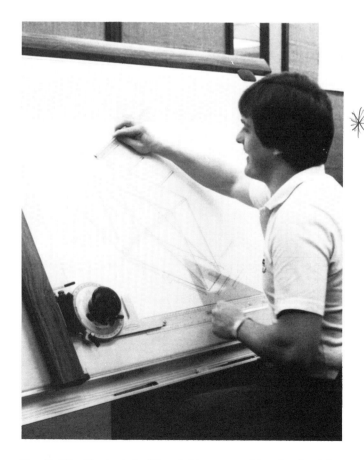

Fig. 2-10. Typical drafting table or machine is tilted for easy work.

DRAFTING KIT

Many drafting supply companies will supply the most common drafting items in a kit. Buying a drafting kit will normally save money. A typical kit will have the items shown in Fig. 2-11.

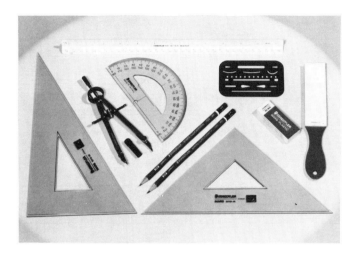

Fig. 2-11. A standard drafting kit.   (Keuffel & Esser Co.)

DRAWING FORMATS

Drawings created in industry must be done on standard sheet sizes. These sizes will be referred to by letter designation, Fig. 2-12. The reasons for keeping all of our documents on standard sized paper are:
1. Drawers and file cabinets are designed around standard sheet sizes.
2. They will fit or fold into a standard book form.
3. They will be the size of standard reproduction paper.
4. The sizes are readily available from suppliers.
   Format sizes larger than ''E'' are designated ''J'' and will be drawn on rolled stock. The maximum recommended length of a roll drawing is 144 in. (12 ft.).

BASIC DRAFTING TECHNIQUES

In order to do a drafting task in a realistic length of time, you must understand some principles. This section will cover some needed drafting techniques.

GEOMETRIC CONSTRUCTION

Geometric construction is a graphical solution to a mathematical problem. The principles of geometry are applied often in engineering. To solve geometric

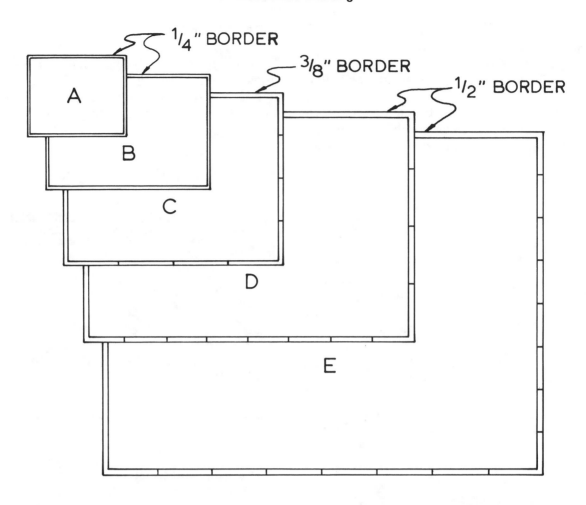

Fig. 2-12. Standard drawing format sizes. Allow for margins plus some extra.

STANDARD FLAT SHEET SIZES

LETTER	WIDTH	LENGTH OR	WIDTH	LENGTH
A	8.5	11	9	12
B	11	17	12	18
C	17	22	18	24
D	22	34	24	36
E	34	44	36	48

problems, you will normally use a straightedge, triangles, compass, scales, and french curve. Below are some common problems worked for you:

1. Draw parallel lines using a triangle and a straightedge, Fig. 2-13.

Sample Problem: Add lines C and D.
   a. Keeping the triangle firmly against the base straightedge, align its edge to B.
   b. Slide the triangle to C and draw the parallel line.
   c. Slide the triangle to D and draw the parallel line.

2. Bisect a line or area using a straightedge and triangle, Fig. 2-14.

Sample Problem: Add a line to the center of the electronics component.
   a. Align straightedge parallel to component.
   b. Using a triangle on the base, draw a line from left corner, (1).
   c. Flip triangle over and draw from right corner, (2).
   d. Slide the triangle over and draw a perpendicular through the intersection to the component. The line will bisect the component.

3. Draw perpendicular lines using a triangle and straightedge, Fig. 2-15.

Sample Problem: Add line C to the schematic 1/4 in. away from line A.

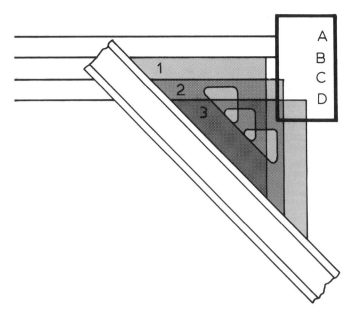

Fig. 2-13. Drawing lines parallel to line A and B.

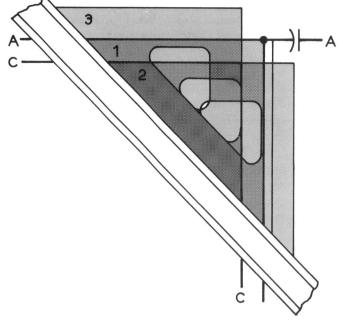

Fig. 2-15. Drawing a line perpendicular to another line.

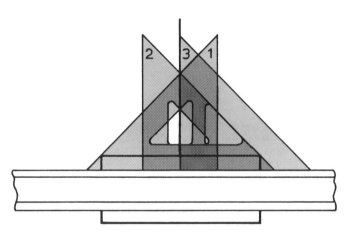

Fig. 2-14. Drawing a bisecting line perpendicular to a given rectangle.

  a. Align the top edge of the triangle to line A, (1).

  b. Slide down 1/4 in. space and draw line, (2).

  c. Slide up to create 1/4 in. vertical space and draw line, (3).

4. Draw a tangent to two circles using a triangle and straightedge. Find the tangent points, Fig. 2-16.

Sample Problem: Draw a transistor.

  a. Draw the two circles.

  b. Align the triangle to the top of both circles, (1).

  c. Slide the triangle up to the center of small circle, (2). Mark the tangent point, (T_1).

  d. Slide the triangle up to center of large circle, (3). Mark tangent point, (T_2).

  e. Draw required tangent line (1).

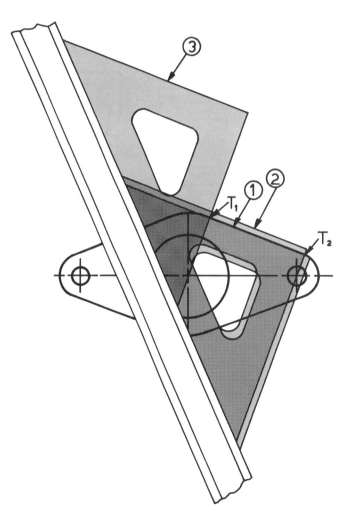

Fig. 2-16. Drawing a line tangent to two circles and locating tangent points.

5. Draw a section of sheet metal rounding a curve. Use a compass and scale, Fig. 2-17.

Sample Problem: Bend sheet metal around 1/2 in. radius. Metal is .12 in. thick.

a. Set compass to make 1/2 in. arc.
b. Set compass at point C and mark off 1/2 in. distance on lines CA and CB.
c. Move compass down line CB to 1/2 in. mark. Draw a new 1/2 in. arc going through O.
d. Move compass down line AC to 1/2 in. mark. Draw a new 1/2 in. arc going through O intersecting arc drawn in step (c). This will be the center of the sheet metal's radius.
e. Set compass at O and draw a 1/2 in. arc striking lines CA and CB at 1/2 in. tangent points.

f. Set compass to make a .62 in. radius (.50 + .12).
g. Locate compass at O and swing an arc parallel to 1/2 in. arc.
h. Connect straight lines at tangent points.

6. Divide a space into a number of equal parts and draw parallel lines using a scale, triangles, and straightedge, Fig. 2-18.

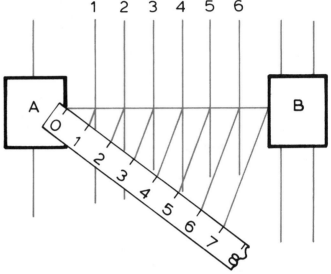

Fig. 2-18. Dividing an area into equal units.

Sample Problem: Draw six equally spaced parallel lines between component A and component B.

a. Draw a line between A and B.
b. Set scale with zero on component A at line.
c. At any angle, lay off seven equal convenient spaces (to obtain six equally spaced lines).
d. Connect the seventh measurement to B at line.
e. Draw six lines parallel to line 7-B to intersect line drawn between A and B.
f. Draw new vertical lines through intersections parallel to existing lines.

7. Draw a hexagon with a 1/2 in. distance across the flats. Use a compass, 30-60 triangle, and straightedge, Fig. 2-19.

Sample Problem: Draw a hexagonal fastener with a 1/2 in. measurement across the flats.

a. Set the compass to measure 1/4 in.
b. Draw a 1/2 in. diameter circle.
c. Construct vertical and horizontal center lines through the circle center point.
d. Set straightedge parallel to horizontal center line.

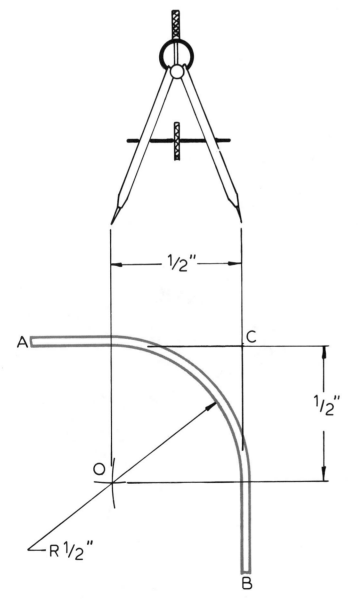

Fig. 2-17. Drawing a section of sheet metal rounding a curve.

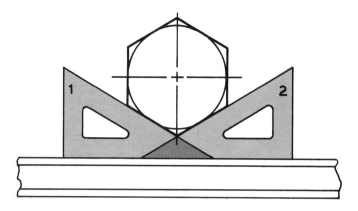

Fig. 2-19. To draw a hexagon, use 30° angles.

e. With the 30-60 triangle, construct tangent lines to outside of circle.
f. Draw vertical lines to complete hexagon.

ORTHOGRAPHIC DRAWINGS

The term orthographic projection means: Projection in which the projecting lines are perpendicular to the image shown, Fig. 2-20. Orthographic drawings are normally two or more views of an object. Each view is of a different side. Orthographic drawings have six principal views. You must learn to choose which ones are important to the reading of the drawing. See Fig. 2-21. At times, the six principle views will not show the object adequately. To show an inclined surface or feature more clearly, you will need to draw an auxiliary view.

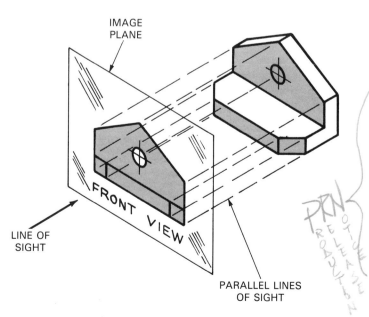

Fig. 2-20. Each line projected from the object will strike the image plane at a perpendicular angle. Note: All the lines are parallel to each other.

AUXILIARY VIEWS

Auxiliary surfaces are those which appear as inclined surfaces in normal views. Auxiliary views are used for determining true size and shape of a surface or angle. A plane's true shape can only be seen when the direction of view is perpendicular to the plane, Fig. 2-22.

SECTIONAL VIEWS

When a part's interior features are too complicated to explain with hidden lines, sectional views are used. A sectional view is created by cutting away material to reveal the interior features, Fig. 2-23. In electronics assemblies, we often need section cuts through the chassis to show how the assembled parts go together, Fig. 2-24. Sectional views, then, serve the functions of:
1. Adding clarity.
2. Helping in assembly of parts.

DRAWING TITLE BLOCKS

Most companies will have their own title block. Fig. 2-25 shows the general information given in the title block. Let's examine the parts drafters fill in. Follow the number labels in Fig. 2-25.
1. Most companies will require the drafters to letter their names and date of completion.
2. The checker will sign in this space and date the completion of the check. The checker will sign only after all the drawing's information has been checked.
3. The engineer in charge will sign and date the approval block. The engineer's signature will verify that the drawing is correct and ready for release.
4. This block is for special remarks.
5. Preliminary release will be signed by the engineering supervisor. This signature will allow copies of the drawing to be distributed to manufacturing and purchasing.
6. Final drawing release will be signed by the engineering supervisor only after the signatures have been made in block 7 and 8.
7. Manufacturing will sign this block. The signature will state that the item can be manufactured and that the design is economically feasible.
8. Purchasing must be able to buy the material, hardware, or components. If the purchasing officials see a problem, they will not sign until they have conferred with the engineers.
9. Reproduction processes will not create an exact scale of the drafter's work, so no attempt

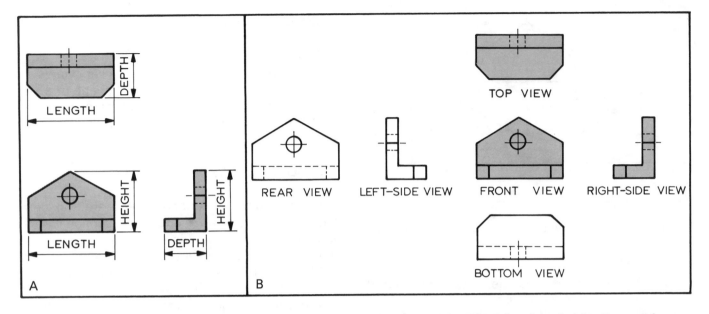

Fig. 2-21. A—This is a three view orthographic drawing. The front and right side views show height. Top and front show length. Top and right side views show thickness. B—The six primary orthographic views. Note: The left side, bottom, and rear views add nothing to the content. They should not be drawn.

should be made to measure the blueline prints or other reproductions.

10. The company's name and address will fill this block.

11. The title of the drawing will be lettered here. Room is provided for three lines in the title.

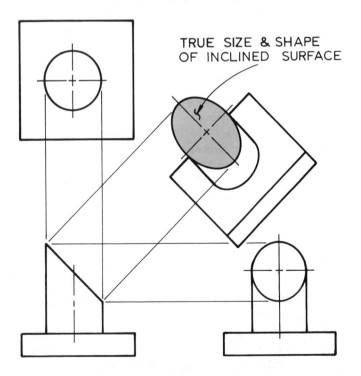

TRUE SIZE & SHAPE OF INCLINED SURFACE

Fig. 2-22. An auxiliary projection drawing. To see the true size and shape of a surface, project 90° from it to the image plane.

12. The size of the original drawing format will be given here.

13. Each company will have a code number. This number will normally be five digits. These numbers are listed in the Handbook for Federal Supply Code for manufacturers, name and code.

14. The drawing number will come from a Drawing Title and Number book.

15. Each time a drawing is changed, it is given a drawing revision letter. The first change will be identified "A." Each successive change will use the next letter of the alphabet, except that "I," "O," "Q," "X," are not used.

16. The general scale of the drawing is given here. If sections or details are drawn at different scales, they will be indicated on the drawing.

17. The unit weight will be written only when required.

18. Enter sheet one of total amount on *sheet one* of multisheet drawings. Additional sheets are numbered in sequence as follows: "sheet 2 of 4," etc.

TYPES OF DRAWINGS

The electronics drafter will create many kinds of drawings. These can be mechanical drawings as well as electronics drawings. A summary of the common types is given in Fig. 2-26. Some of the drawing types are used for quality control, manufacturing, research and development, or marketing.

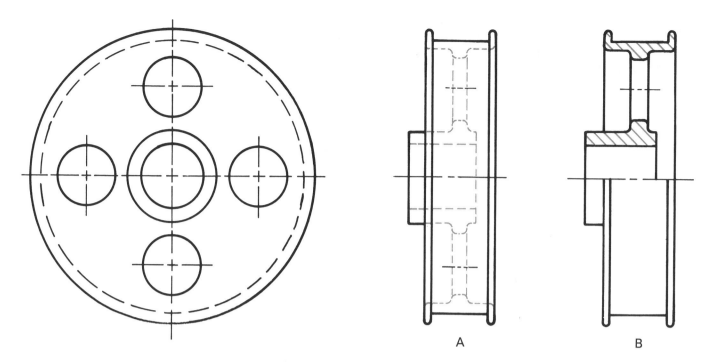

Fig. 2-23. A pulley with two right side views. A—Hidden lines describe the interior features. B—Material is cut away to show the interior features. View B is a half section.

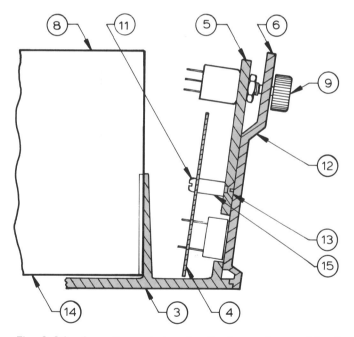

Fig. 2-24. A section cut showing the internal assembly of an electronic machine.

This book discusses only a few of the subjects in detail. You may wish to consult specialized drafting books for more information.

COMPUTER-AIDED DESIGN

Computer-aided design (CAD) was pioneered in the electronics and the aerospace industries. CAD allows the drafter to create, store, and alter

DRAWN	①	/ /	⑩				
CHECKED	②	/ /					
APPROVED	③	/ /	TITLE				
	④		⑪				
PREL.	⑤	/ /					
FINAL	⑥	/ /	SIZE	IDENT. NO.	DRAWING NO.		REV.
MAN.	⑦	/ /	⑫	⑬	⑭		⑮
PURCHASE	⑧	/ /					
⑨ DO NOT SCALE DWG.			SCALE ⑯	UNIT WT: ⑰		SHEET ⑱ OF	

Fig. 2-25. A typical drawing title block.

Mechanical or Electrical	Electronics
Layout	Schematic
Monodetail	Connection or wiring
Multidetail	Wiring harness
Tabulated	Cable assembly
Assembly	Running wire list
Detail assembly	Interconnection
Photographic assembly	Single-line
Inseparable assembly	Logic
Arrangement	Printed wiring
Specification control	Block
Source control	Flow
Installation	
Mechanical schematic	
Book-form	
Numerical control	
Outline	
Standard	
Graphic or chart	

Fig. 2-26. Types of drawings needed for quality control, design, or production.

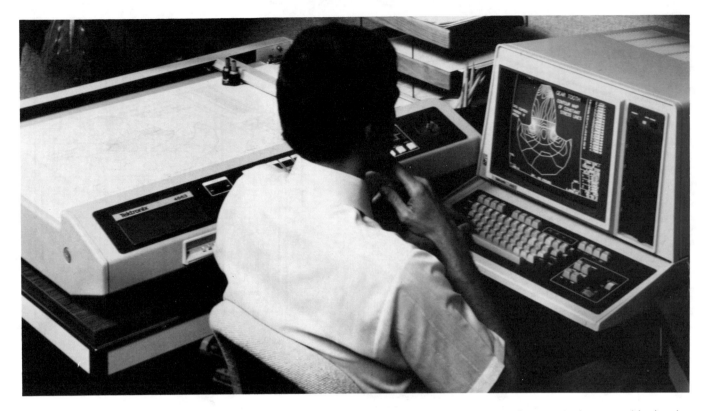

Fig. 2-27. New designs can be tried on a computer-aided design system. Note plotter used to provide hard copy. (Tektronix, Inc.)

engineering drawings using a computer, Fig. 2-27. This technological advancement has the potential to change and improve industrial productivity as much as any technological development since electricity.

Computer-aided manufacturing (CAM) is the logical extension of CAD. CAD/CAM systems have been successful in providing solutions to some of industry's oldest problems. One of the main problems has been the historical separation of engineering and manufacturing. This separation causes a breakdown in communication, and shows the transfer of data between engineering and manufacturing.

Communication suffers when drawings are sent from engineering to manufacturing with inadequate or incomplete dimensioning, omitted part callouts, incorrect materials listed, and other errors. The computer helps in eliminating these costly errors.

Poor data causes many problems. Obsolete drawings used in manufacturing result in rejected parts. This causes a waste of time and materials. To solve this problem in a traditional way, we strive to get the latest blueline prints to manufacturing immediately after completing the drafting work. Drawing management does not always make this transfer of data in a timely manner. CAD/CAM can aid management with these problems.

CAD/CAM can help management with the engineering and manufacturing communications problem. Both departments have computer terminals which are linked to the same data bank. This allows manufacturing to use the computer to verify the latest drawing update. Manufactuirng can, with the aid of the computer, instantly call up for review any print it desires, saving valuable time and storage space.

CAD/CAM FUNCTIONS

The main purpose of CAD/CAM is to relieve the drafter and assembler of the boredom of doing repetitive tasks. When you have common tasks to perform, you can program the computer to do them for you.

The computer can also be used as an analytical tool. It can save you hours of computation time because it can verify designs and functions. It avoids the old production by trial and error. Fig. 2-28 shows how CAD/CAM works on a circuit board process.

The computer can control harness assemblies, panel assemblies, flow diagrams, and most electronic drawings. This work is possible because of the repetitive nature of these drawings.

With printed circuits and integrated semiconductor circuits, the demand for size and increased drafting output makes the computer system necessary. A drafter cannot manually match the accuracy and speed of the computer when doing high density printed circuit design.

CAD has some advantages for the drafter. These advantages are:

1. Standard symbols or objects can be stored by the computer. Then as the drafter requires a symbol or object, the symbol can simply be "CALLED UP," saving valuable time required to draw it again. "Call up" means to recall it from the computer's memory. The computer will also copy any feature or group of features you have on a drawing.

2. It allows a quick and easy storage file. Once a drawing has been completed. It can be saved in the computer's storage system. Later, you can call up the drawing for design changes, analysis, or other engineering requirements. Storage files are normally magnetic tape or disk recordings.

3. Drawings can be created more quickly and accurately. The more complex the drawing, the greater the time savings. Some companies re-quire the computer to save 80% of their drafting costs. If the computer cannot do this, the company may choose traditional drafting methods.

4. The CAD drawing can be produced to ±.001 in. accuracy, depending on the plotter. Traditional drafting methods will not allow for a timely completion of a drawing to this accuracy.

5. Drawing changes can be made much easier. Areas of the drawing can be deleted and new detail added without laborious erasing and redrawing.

6. When combined with CAM, it becomes a very powerful tool for both engineering and manufacturing. An example is in printed circuit board design. The CAD information is used in all steps through the final inspection of the completed board.

7. Prototype drawings can be quickly altered to show many design possibilities. Parts can be moved to show possible interferences or design errors. Mating parts can be drawn on different layers so that you can display one, or both, at the same time to check their mating features.

8. Automatic parts lists can be created by placing a hidden parts number below the symbol. When

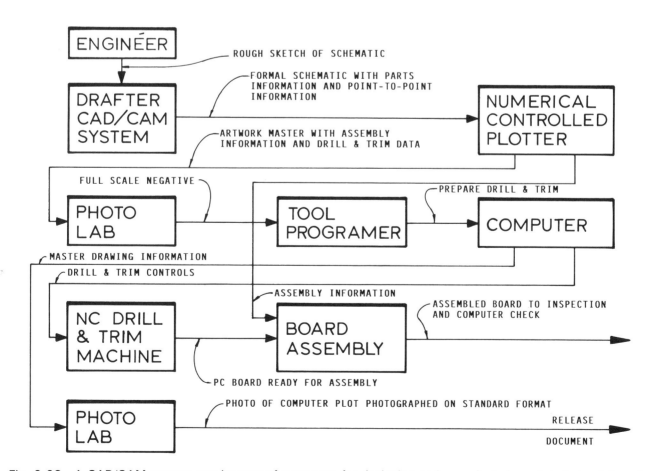

Fig. 2-28. A CAD/CAM system can be part of a process for designing and manufacturing printed circuit boards.

a resistor of a specific value is added to a drawing, an invisible part number is also added. After the drawing is completed, the computer can scan the drawing, and list the following:

a. Part number.
b. Quantity of parts.
c. Parts description.

The computer can compile large parts lists in minutes.

After reading this list of existing possibilities, you may ask, "Will computers replace drafters?" Most companies say no. They see the drafter using the computer as an added tool. The computer will allow the drafter to do more engineering design, data transfer, and pattern drawing control.

Drafters will be needed to generate data banks. As new engineering data is generated, the banks of data will need to be updated. Besides using this data to create graphics, the drafter will be more in touch with management. The data bank established by the drafter will be needed for management control. For example, purchasing, production, and accounting will all need drafting-generated computer information.

COMPUTER SYSTEM PARTS

In addition to software, the computer requires hardware in order to be useful graphics tool. Some of the hardware parts are described in the following list.

CENTRAL PROCESSING UNIT (CPU)

The CPU is the brain of the system. It works with the data, and it executes the commands. The CPU also contains the computer's temporary memory.

MONITOR

The monitor displays the communication between you and the computer, and vice-versa. The monitor also works as a graphics display.

PLOTTER

The plotter creates a HARD COPY (paper copy), as opposed to the SOFT COPY (electronic image) the monitor produces, of the data. Soft copy is lost when the power is turned off.

KEYBOARD

The keyboard is one way to input data to the computer. It is similar to a typewriter keyboard, and it can easily be operated by a trained typist. The drafter can make alphanumeric inputs on a normal keyboard, but many computers have additional function keys.

MOUSE

A mouse is an effective way to input graphic data to the computer. It allows information to be entered easily by moving the hand-held mouse across a flat surface.

MODEM

The word modem stands for modulator/demodulator. The modem is a telephone linkage unit. It connects a computer to a data source or to another computer over the telephone lines.

DISK DRIVE

A disk drive is an I/O (input/output) devise. It can feed the computer prerecorded information or can be used for storage of the output data.

COMPUTERS IN ELECTRONIC COMPANIES

The CAD systems have chiefly been used to lay out printed circuit boards and integrated semiconductor circuits. But they are now being used to analyze circuits. This is done at different levels of completion:

1. When the engineer finishes the design, the computer can help check it.
2. It will check the printed circuit artwork. It checks completeness and accuracy of the artwork.
3. A final check on the assembled circuit board will disclose any bad components or board errors.

Another area where the computer is aiding design is in mechanical packaging. This is accomplished by modeling the parts or package.

REVIEW QUESTIONS

1. Why should electronics drawings be completed with high quality?
2. What requirements mandate the 5/32 in. lettering?
3. List the reasons for standardization in drawing sheet sizes.
4. How many primary views are there in orthographic projection?
5. Auxiliary views serve what main purpose?
6. What view will be drawn when we need to see internal features?
7. When will lower case lettering normally be used?

8. Drawings larger than ''E'' size are _____ sized format.
9. Why are thin lead pencils becoming popular in the drafting field?
10. To draw long pencil lines that are sharp, you should:
   a. Stop often to sharpen the pencil.
   b. Use three or more pencils in succession.
   c. Use a vary hard lead in the holder.
   d. Rotate the pencil while drawing.
   e. Statements c and d above.
11. Make two or more parallel lines using a straightedge and a _____.
12. What potential will the computer have in industry?
13. CAM means:
   a. Computer Assembly and Maintenance.
   b. Computer-Aided Machining.
   c. Computer-Aided Manufacturing.
   d. Computing and Machining.
14. A CAD system should save at least _____% of drafting costs.
15. List some advantages of CAD.
16. Will the computer replace the drafter? Explain.
17. What is hard copy and soft copy?

PROBLEMS

PROB. 2-1. Letter the alphabet using gothic upper case vertical letters. Your instructor will establish the requirements.

PROB. 2-2. Letter freehand the following information using upper case vertical lettering and guide lines:
   NOTE:
   1. UNLESS OTHERWISE SPECIFIED: RESITANCE VALUES ARE IN OHMS, 1/4W, and 10% CAPACITANCE VALUES ARE IN MICRO-FARADS, 35V, and 5%
   2. SOLDER ALL TERMINATIONS PER TAYLOR WORKMANSHIP STAN—DARDS
   3. THIS DRAWING TO BE USED IN CONJUNCTION WITH ASSEMBLY DRAWING 780629-5

PROB. 2-3. Letter the alphabet using lower case lettering. Your instructor will establish the requirements.

PROB. 2-4. On a separate paper, complete the geometric exercises in Fig. 2-29.

PROB. 2-5. Draw the necessary orthographic views of the part shown in Fig. 2-30.

PROB. 2-6. On a separate sheet, draw the two existing views and an auxiliary view of the part which is shown in Fig. 2-31.

Draw three lines with equal spacing and parallelism between them.

Bisect a 2.50 line.

Draw a perpendicular to a given line.

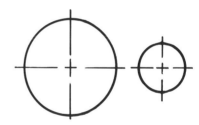

Construct a tangent line between a 2.00 dia. and 1.00 dia. circle spaced on 2.00 inch centers.

Divide a 6 1/4 in. line into 6 equal parts.

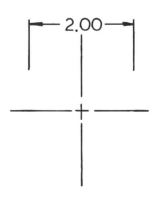

Draw a hexagon shape with a 2.00 in. measurement across the flats.

Fig. 2-29. Draw the above problems using techniques shown earlier in the chapter.

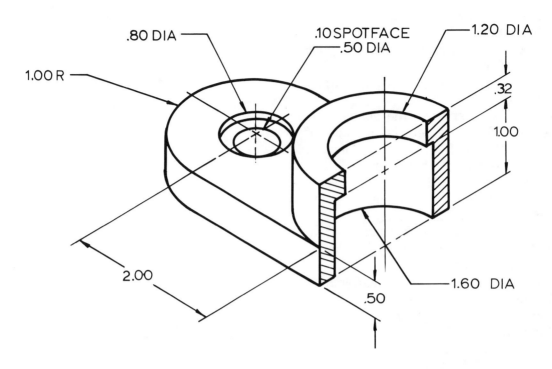

Fig. 2-30. Draw the necessary orthographic views of the part.

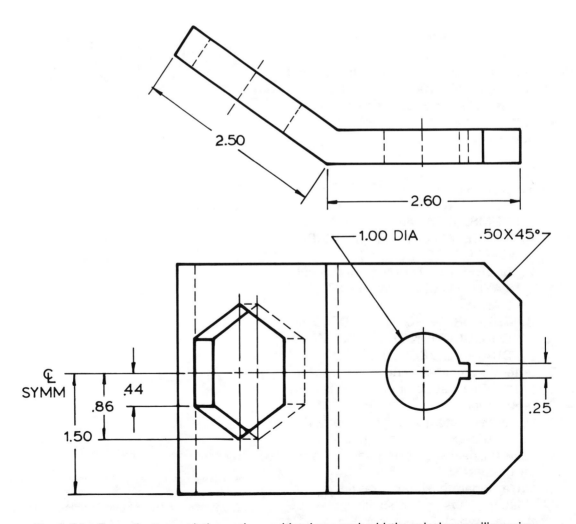

Fig. 2-31. Draw the two existing orthographic views and add the missing auxiliary view.

Chapter 3

BLOCK, FLOW, AND SINGLE LINE DIAGRAMS

After studying this chapter, you will be able to:
- ☐ Draw a block diagram, flow diagram, and a single line diagram.
- ☐ List the rules for a correctly drawn block, flow, and single line diagram.

Electronics systems are created from ideas. To advance these ideas, we must look at them on paper. Putting our thoughts on paper will help us organize ideas and check if our ideas are workable.

The arrangement of our ideas on paper is called a diagram. You are going to study three different types of diagrams in this chapter.

BLOCK DIAGRAMS

Block diagrams are the most elementary of all the electronic drawings. These diagrams show only the essential units of the system. The units are normally represented by rectangular blocks. See Fig. 3-1.

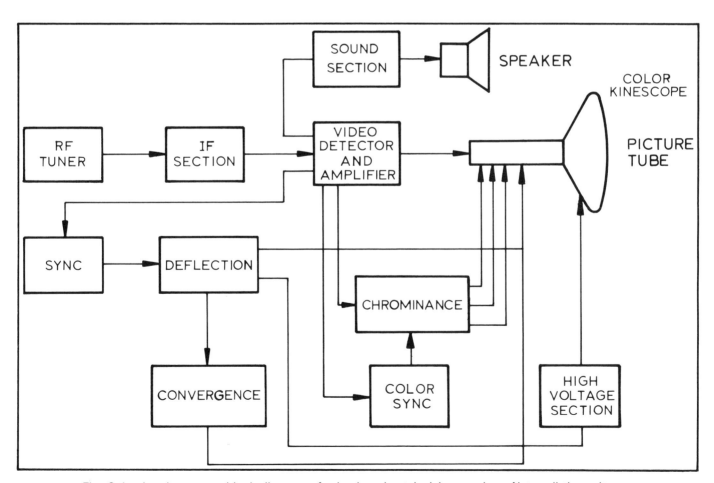

Fig. 3-1. An elementary block diagram of a basic color television receiver. Note: all the units are shown as rectangles except for the speaker and picture tube which are shown by symbol.

The blocks are tied together with lines. The path of the signal or energy may be shown by the lines or by arrows. Arrows show its direction. Block diagrams are used by computer programmers to aid in program development. This diagram will not show graphical symbols or reference designations.

A simplified block diagram of a television receiver would look like Fig. 3-2. Take a look at what the diagram accomplishes.

The engineer will sketch a block diagram of the television system. Using this diagram he/she will help decide what each of the units is supposed to do. The engineer established the following requirements: (As you look at Fig. 3-2, match each number with its unit.)

1. Low voltage power supply. This unit will convert the 120 volt alternating current to the appropriate direct current voltage levels.
2. An RF tuner. The radio-frequency (RF) tuner will need to select the desired channel and reject all others. It will be designed so that the signal will also be amplified.
3. An IF amplifier. The intermediate-frequency (IF) amplifier will provide additional amplification for the video and sound signals.

4. A video detector and amplifier. This section will set up the picture and demodulate (decode) the IF signal.
5. A sound section. This is where the audio amplification and power output takes place.
6. Synchronization section. This section will separate the horizontal and vertical pulses and route them to their respective deflection sections. It must keep the pulses synchronized as it transmits them.
7. Deflection section. The horizontal and vertical movement of the electron beam is provided by this unit.
8. High-voltage section. This section will set up the 15-20KV source which is necessary to develop a sharp, bright picture.

When the engineer has the basics for the television set figured out, he/she will assign the project to the drafter.

Block diagrams are used in sales literature, service manuals, electronics catalogs, and operator manuals. Compared to other diagram types, the block diagram is understood better by persons unfamiliar with electronics. It shows the functional relationship of each stage in the simplest way.

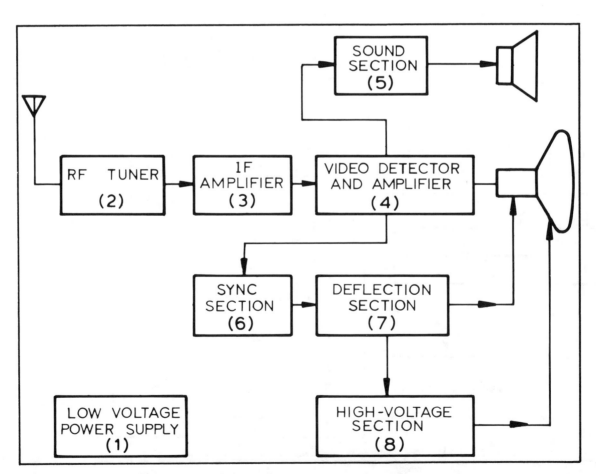

Fig. 3-2. A typical block diagram for a simple television receiver. Numbers in the blocks will correspond to written explanation in the paragraph on block diagrams.

RULES FOR DRAFTING BLOCK DIAGRAMS:

1. Blocks within a diagram should normally be the same size, Fig. 3-3.
2. The block containing the most information will establish the block size for the drawing, as shown in Fig. 3-3.
3. Inputs to the rectangles should come in the left side or top of the block. See Figs. 3-4 and 3-5.
4. Outputs should go out the right side or bottom of the block, Figs. 3-4 and 3-5.
5. Interconnection lines must run horizontally or vertically with all corners at 90 degrees, as shown in Fig. 3-6.
6. Lines running parallel to each other should be grouped with a larger space between every third line. This helps the reader's eye follow each line. See Fig. 3-7.
7. Minimize crossed lines, Fig. 3-8.
8. Minimize jogged lines, Fig. 3-8.
9. If symbols are used, they should be from ANSI Y32-2, Fig. 3-9.
10. Lettering between elements of the system should either be done above, above and below, below, or in the line, Fig. 3-10.

Block diagrams and flow diagrams use basically the same rules. Let's look at flow diagrams to see their similarities and differences.

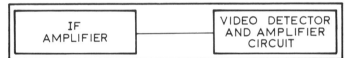

Fig. 3-3. One of eight examples of block diagrams and design rules. This demonstrates rule 1 and 2.

Fig. 3-4. An example of rule 3 and 4.

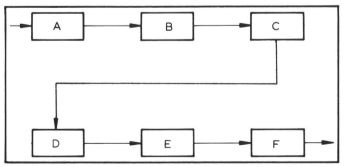

Fig. 3-5. An example of rule 3 and 4 when paper size dictates layout.

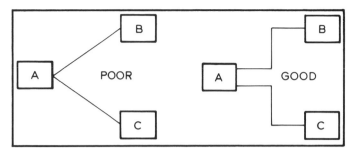

Fig. 3-6. An example of rule 5.

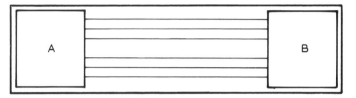

Fig. 3-7. An example of rule 6.

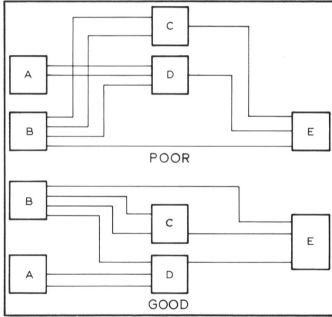

Fig. 3-8. An example of rule 7 and 8.

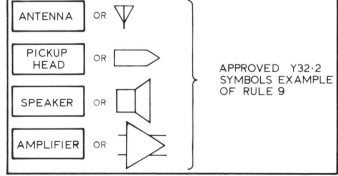

Fig. 3-9. An example of rule 9.

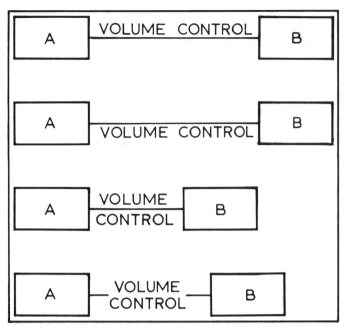

Fig. 3-10. An example of rule 10.

FLOW DIAGRAMS

Flow diagrams are sometimes called flowcharts. Flow diagrams show the sequence of events in a process or operation. The sequence will normally start at the top of the sheet and go down, Fig. 3-11. Sometimes the block diagram will go left to right horizontally. The logical steps of a computer program can be neatly shown by the flow diagram, Fig. 3-12. A program is coded by writing down the successive instructions that will cause the computer to perform logical operations.

RULES FOR DRAFTING FLOW DIAGRAMS

1. Sequence is from top to bottom.
2. Draw all boxes same width.
3. Box height can vary.
4. Use thick lines on boxes, medium lines on arrows.

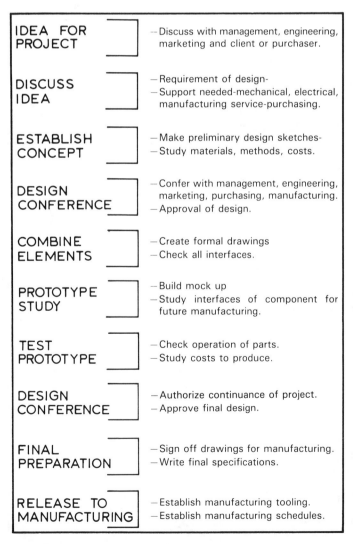

Fig. 3-11. A typical flow diagram has operation sequence from top to bottom of page.

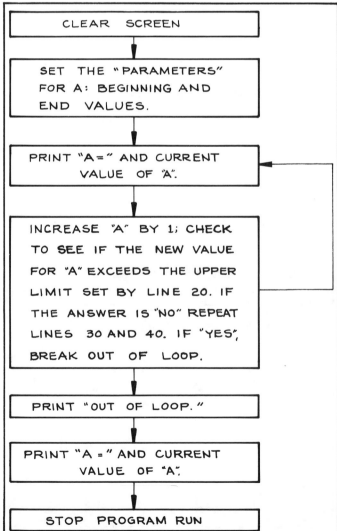

Fig. 3-12. A flow diagram of a for-next computer program.

5. For arrows going back to earlier steps, make side spacing at least 1/4 the width of a box.
6. Use hand lettering within guidelines.

SINGLE LINE DIAGRAMS

The next step following a block diagram, may be a single line diagram. See Fig. 3-13. This diagram is a form of schematic, which uses a single line to show component interconnections. The single line may represent many lines in the actual circuit. It will omit the detailed information shown on schematics or connection diagrams.

Single line diagrams will show:
1. The relationships between circuits.
2. Meters, instruments, switches, relays, and other power circuit devices.
3. Ratings of circuits which are essential to the overall understanding. Generator ratings for example, include: kilowatt capacity, power factors, voltage, cycles, revolutions per minute, and number of windings.
4. Neutral and ground connections.
5. Feeder circuits.
6. The general layout of the circuit. The layout will show only the information needed for clarity.

RULES FOR DRAFTING SINGLE LINE DIAGRAMS:

1. Line thickness may vary, Fig. 3-13. Thick lines indicate primary information lines, medium lines indicate connection to power source.
2. Circles are used to portray meters, motors, instruments, and other rotary equipment.
3. Rectangles depict resistors, switches, components, and other major equipment parts.

The single line diagram is the closest drawing to the schematic. Schematic drawings will be covered in chapter five. Chapter four will cover the individual electronic symbols and reference designations. We must have an understanding of the components, their symbols, and references before continuing our study.

REVIEW QUESTIONS

1. What does a block diagram show?
2. How can we show the path and direction of the signal or energy?
3. List the major differences between a block and flow diagram.
4. List what a single line diagram shows.
5. Flow diagrams may also be called _____.

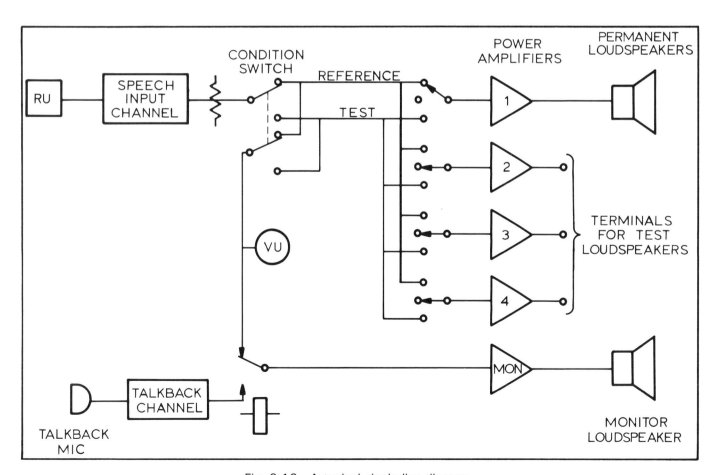

Fig. 3-13. A typical single line diagram.

6. Why is it important to put our ideas on paper?
7. List where block diagrams are used.
8. What does a flow diagram show?
9. What is the closest diagram to the schematic?
10. Why do we group lines with a larger space between every third line when they are running parallel?
11. The type of diagram least understood by those untrained in electronics is the:
    a. Flow diagram.
    b. Single line diagram.
    c. Block diagram.
12. If arrows in a flow diagram go back to previous steps, make the spacing at the side of the chart at least _____ the width of a box.
13. Draw boxes in a flowchart with _____ (thicker, thinner) lines than for arrow lines.

PROBLEMS

PROB. 3-1. Draw a block diagram from information given in Fig. 3-14. Follow the rules for a properly drawn block diagram.

PROB. 3-2. Draw a block diagram of Fig. 3-15. Follow the 10 rules.

PROB. 3-3. Draw a single line diagram of Fig. 3-16.

PROB. 3-4. Draw a flow diagram of Fig. 3-17.

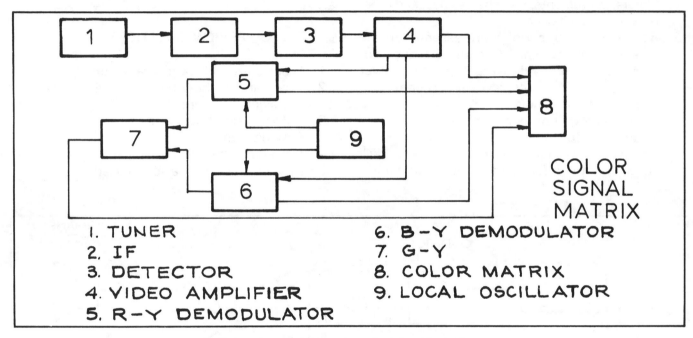

1. TUNER
2. IF
3. DETECTOR
4. VIDEO AMPLIFIER
5. R—Y DEMODULATOR
6. B—Y DEMODULATOR
7. G—Y
8. COLOR MATRIX
9. LOCAL OSCILLATOR

Fig. 3-14. Redraw this color television block diagram.

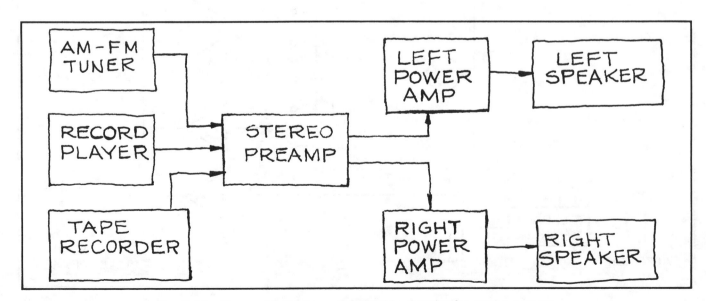

Fig. 3-15. Draw a block diagram from this sketch of a stereo system.

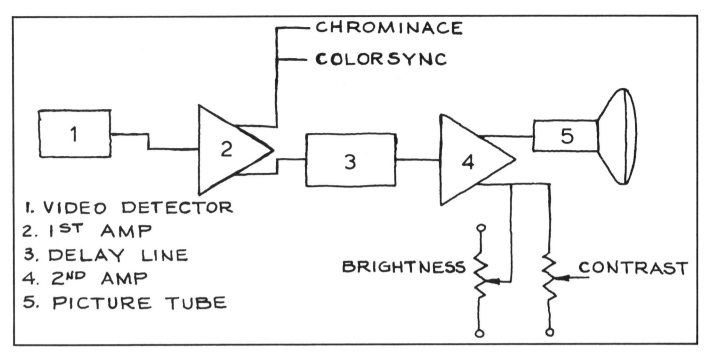

Fig. 3-16. Draw a single line diagram of this television amplifier system.

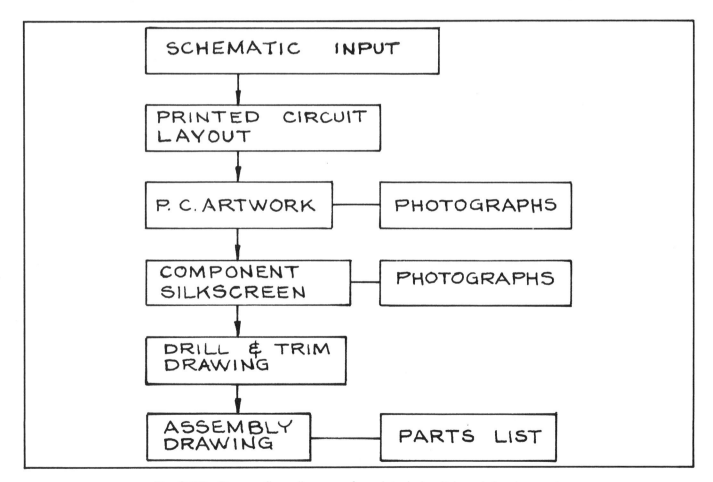

Fig. 3-17. Draw a flow diagram of a printed circuit board development.

Chapter 4

ELECTRONICS SYMBOLS, COMPONENTS, AND REFERENCES

After studying this chapter, you will be able to:
- Identify components by symbol.
- Read the resistor color code.
- Correctly draw component symbols with a template.
- Correctly reference components.
- Correctly write component values.

Electronic circuits are normally made up of individual components. The drafter's knowledge of these components, their symbols, and references is mandatory. You need to know these important facts so you can represent the components in a schematic. The engineer will design the circuit and analyze its feasibility.

After completing the engineering task, a sketch of the circuit will be submitted to drafting. Drafting will use the sketch to create a formal schematic drawing. The drafting department is responsible for making sure each component is correctly shown. To do this, you need to be familiar with the following standards:

1. Y32.2 ELECTRICAL AND ELECTRONICS DIAGRAMS, GRAPHIC SYMBOLS for.
2. Y32.14 LOGIC DIAGRAMS, GRAPHIC SYMBOLS for
3. Y32.16 REFERENCE DESIGNATIONS for ELECTRICAL AND ELECTRONICS PARTS AND EQUIPMENT.

These standards will assure that your drawings are correct and have industry-wide acceptance.

RELATIONSHIP OF COMPONENTS AND SYMBOLS

In many cases the symbol very closely resembles the physical component. The switch is a good example. Note the relationship in Fig. 4-1. In studying this chapter, look for other symbols which closely resemble their components.

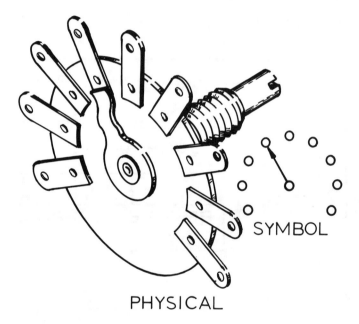

Fig. 4-1. A rotary switch and symbolic representation.

COMPONENTS

There are many different components used in electronics. The scope of this book will allow you to study only the basic ones. You will start with the resistor.

RESISTOR

The resistor is a component which introduces a specific RESISTANCE into the circuit. See Fig. 4-2. Resistance is the opposing of electron flow. The amount of opposition is regulated by changing the length, diameter, or material of the conductor. Resistors are normally made of carbon or nicrome wire. Both of these materials are poor conductors of electricity.

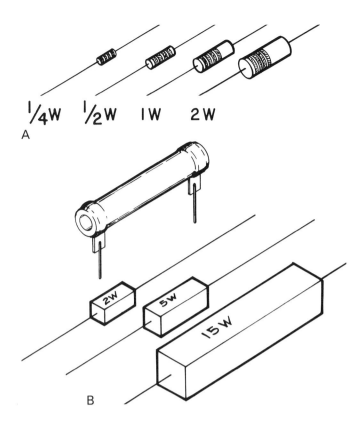

Fig. 4-2. Some typical resistor styles. A—Fixed carbon resistors sized by wattage rating. B—Fixed, wirewound, high temperature resistors with power ratings of 2 watts and above.

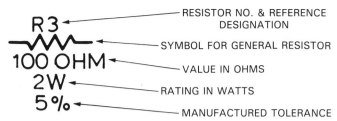

Fig. 4-3. A resistor symbol with complete information.

Resistors are referenced with the letter "R." Each component family will have a different letter to reference it, Fig. 4-3.

Resistors are valued in ohms. Their values can range from a fraction of an ohm to millions of ohms. Carbon resistors have a color code which is used to identify their values (see appendices for resistor color code).

Resistors are also rated in watts. The value in watts is the maximum power the resistor can safely handle. Carbon resistors normally range from 1/8 to 2 watts. Resistors above 2 watts are normallly wirewound. The resistors will be larger as the voltage increases.

Resistors, like other components, cannot be made perfect. A tolerance must be given to allow for manufacturing errors. The tolerance will normally be a 1 to 10 percent variation from the stated value.

GENERAL RESISTOR

The general resistor is one in which there are no options. It serves the function of supplying a set and stated value. These resistors are called fixed resistors. Now let us look at some adjustable resistors.

RHEOSTAT

The rheostat is one of the variable resistors. It has two terminals. A typical use is to dim the light above your dining table. The symbol for a rheostat is shown in Fig. 4-4A. The moving arrow is called the wiper. The wiper moves across the resistor allowing you to adjust the amount of resistance in the circuit.

In Fig. 4-4B you see a dashed line between the two rheostat symbols. This line means ganged or mechanically coupled components. As the shaft adjustment of the component D moves, it adjusts both rheostats simultaneously. Note: In studying this new language, electronics, you will find other components with arrows. See if they are also variable.

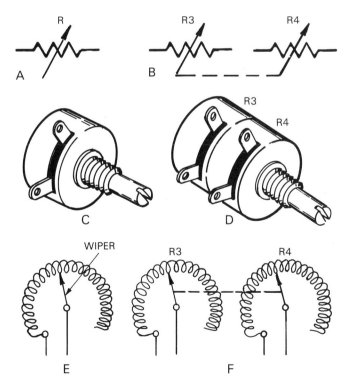

Fig. 4-4. A and B are two symbols used for rheostats. C and D are the physical components. Sketches in E and F suggest how the resistance wire in a rheostat is wound. Clockwise rotation of the wiper increases resistance.

POTENTIOMETER

The potentiometer is also a variable resistor. It is different from the rheostat in that it has three terminals. See Fig. 4-5. It may be used to balance a stereo speaker system.

The potentiometer may also be used as a rheostat. The wiper is tied to one end terminal thus making it a two terminal resistor like the rheostat, Fig. 4-6.

TAPPED RESISTOR

A tapped resistor is normally a wirewound type. See Fig. 4-7. It may have one or more terminals along its length. Tapped resistors are normally used for voltage divider applications.

RESISTOR PACKS

It is possible to purchase resistors packaged together in a single body. This package looks just like an integrated circuit chip, Fig. 4-8. The resistors in the package are normally of the same value.

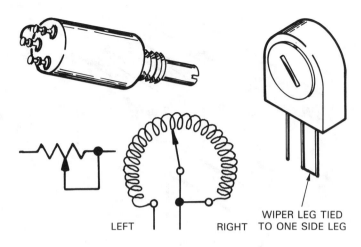

Fig. 4-6. Potentiometers with their wipers tied to one side perform as rheostats.

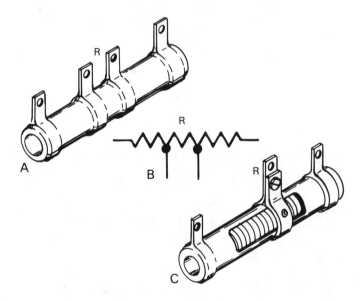

Fig. 4-7. A—A double-tapped resistor. B—The symbol for a double tapped resistor. C—An adjustable tap resistor.

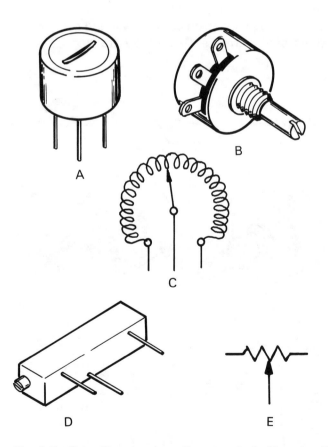

Fig. 4-5. Potentiometers have three terminals. Note the different physical shapes of the components. This is based on how they will be used and adjusted in the equipment. A—Rotary. B—Rotary. C—Symbol. D—Slide. E—Schematic example.

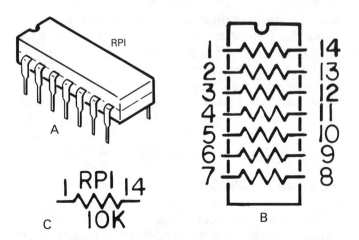

Fig. 4-8. A—One type of resistor package. B—A schematic of the package. C—How to call out a resistor from resistor pack 1.

38

SEMICONDUCTORS

You will be studying a family of components called semiconductors. As components go, semiconductors are relatively new. These are the components that brought about the miniaturization of electronics components. Begin with the diode.

DIODE

The diode is a two-electrode semiconductor. It permits easy flow of electrons in only one direction. The flow is from the cathode to the anode, Fig. 4-9. It is necessary for the drafter to know the cathode and anode ends of the diode. This knowledge will help us show it correctly in the assembly of the circuit.

Note the 1N662 number shown in Fig. 4-9. This number is a catalog number. The engineer will call out this number to indicate the component wanted in the circuit.

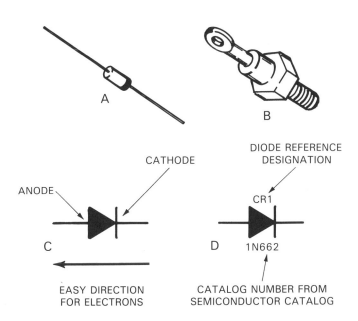

Fig. 4-9. General diode components and symbols. A and B—Typical component shapes. C—Symbol with easy direction shown. D—A symbol with designation (CR) and catalog number.

ZENER DIODE

The zener diode is a breakdown diode, Fig. 4-10. This means it draws more current as the rated voltage is reached. Zeners are used to regulate the voltage of a circuit. They may handle one to hundreds of volts. The zener diode symbol is different from that of a standard diode only in the way the cathode is shown.

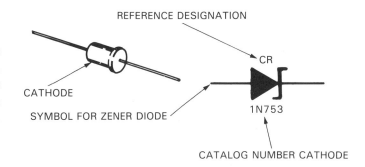

Fig. 4-10. A zener diode symbol.

BRIDGE RECTIFIER

The bridge rectifier is used to convert alternating current into direct current, Fig. 4-11. Alternating current is electrical current that reverses its direction of flow at regular intervals. Direct current is an electrical current flowing in one direction only. A rectifier is used in our cars to change the alternator output into direct current which is needed by the battery and other electrical devices. The bridge rectifier may be called a full-wave rectifier. It has four diodes that work together to allow current in only one direction.

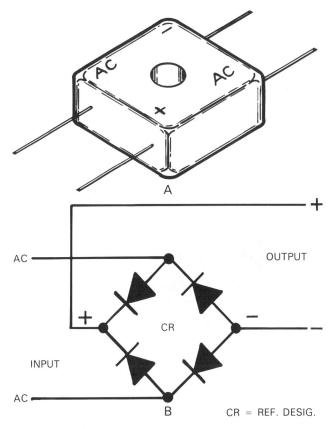

Fig. 4-11. A—A bridge rectifer. B—How the diode elements are linked to perform the rectification.

39

TRANSISTOR

A transistor is an active semiconductor device used in solid state electronics, Fig. 4-12. This component along with the diode has all but eliminated the tube or vacuum tube. It normally has three electrodes: the emitter, base, and collector.

There are two basic types of transistors; the PNP and NPN type. In drawing the symbol the only noticeable difference is the direction of the arrow. The NPN arrow on the emitter points out of the envelope (the circle of the symbol), (A). The PNP has the arrow pointing into the base, (B). A way to remember the NPN type is: "NPN" reminds you that the arrow is "not pointing in." There are other types of transistors, Fig. 4-13. These symbols are for units that perform specialized functions. The symbols will be used less often than for other transistors.

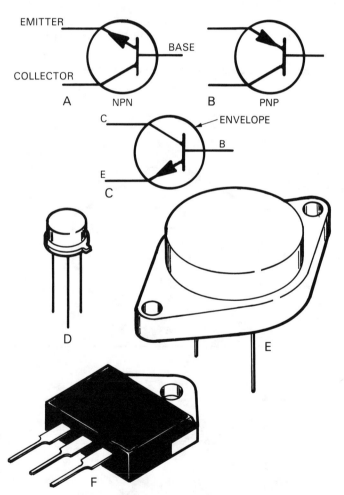

Fig. 4-12. A—An NPN transistor. B—A PNP transistor. C—A transistor symbol with legs identified. D—A transistor case with the right leg identified as the emitter leg. The little tongue is the indicator. E—A transistor which has the body for a collector. E, F—Transistors are both made larger so they can dissipate their heat. Sometimes they are mounted on other metal shapes to help extract their heat.

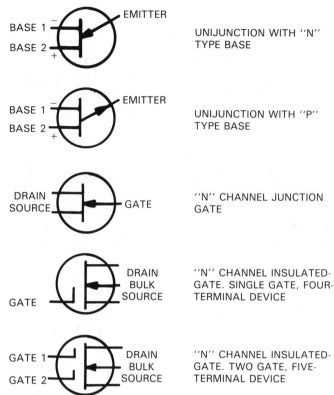

Fig. 4-13. The field effect transistors (FETs) shown in this example have names by their symbols. This is just a book explanation and is not part of the symbol.

INTEGRATED CIRCUIT

An integrated circuit (IC) is an electronics device in which both active and passive components are contained in a single package, Fig. 4-14. These components are electrically interconnected during fabrication. The interconnected parts are then packaged in a protective coating. The package will have flat leads, A, C, or round leads, B, extending to the outside for electrical connections.

Passive components used in IC circuits are resistors, capacitors, and coils. These components are nonpowered and do not create or amplify energy. They rely on a signal to perform their function.

Active components used in IC circuits are transistors and diodes. These components are capable of controlling voltages or currents. They can produce energy or a switching action in the circuit. Their output is dependent on a source of power.

The miniaturization of circuits is one of the most important accomplishments in the field of electronics. The circuits are so small they must be constructed by technicians using microscopes. The circuits are made from very small pieces of silicon, commonly called chips.

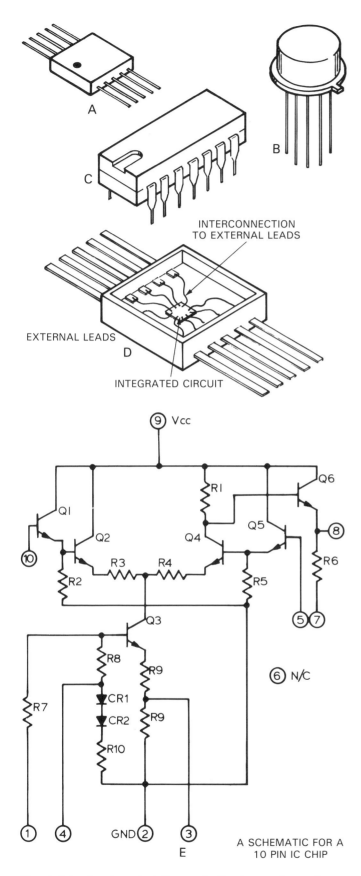

Fig. 4-14. A—A typical flat pack. B—A round metal can package. C—A dual in-line package, the most commonly used style of integrated chip package. D—A flat pack with the internal circuit exposed. E—An example of components normally found inside the IC circuit.

HOW INTEGRATED CIRCUITS ARE MADE

Integrated circuits are created by masking, etching, and diffusion on a MONOLITHIC SUBSTRATE (large sheet foundation) of silicon. The mask is a set pattern used to control selective etching or impregnation of portions of a semiconductor material with impurity atoms. Etching is the removal, by chemicals, of unwanted material from a surface. Diffusion is the process of doping impurities into the silicon to form the desired junctions. From this complex explanation it is evident that a full study of the chip's design and fabrication is beyond the scope of this text. However, we can take a simplified exploration of the chip to give you an appreciation of this device.

Integrated circuits are made on a thin slice of silicon with a diameter of one to two inches. A normal slice may contain 100 to 1000 circuits side by side. After processing, the circuits are separated to make an equal number of individual circuits called chips.

To create a chip, the typical processes are:

1. Get a wafer of P-type silicon as a substrate. The wafer will be thin slice of silicon doped or impregnated with positive impurities, Fig. 4-15.
2. Add a layer of N-type silicon about .20 microns thick. The layer is grown on the wafer. This N-type layer will become the collector for a transistor.
3. Add a thin coat of silicon dioxide. It is grown over the N-type material.
4. Mask the areas to be etched. The mask will establish the areas of acid resist. The wafer is then etched with an acid. The acid resist will cause the desired areas to be left, Fig. 4-16.
5. In the next step, the P-type material is diffused into all the areas not covered by silicon dioxide. Diffusion is the putting upon and into a base a P or N-type material, Fig. 4-17.
6. During the diffusion process, a new layer of silicon dioxide forms over the P-type areas and also on the top of the island.

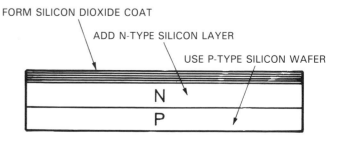

Fig. 4-15. The first three steps in IC construction.

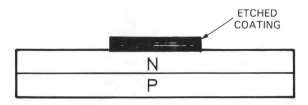

Fig. 4-16. The silicon dioxide layer after etching.

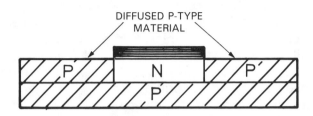

Fig. 4-17. P-type material has been diffused into the unprotected regions.

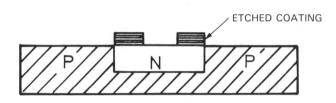

Fig. 4-18. Etching has created an area for a new region.

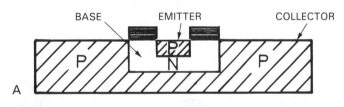

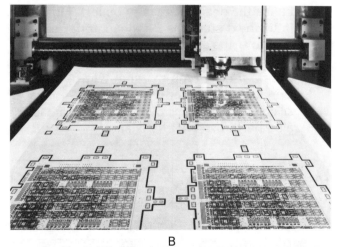

Fig. 4-19. A—The steps have shown how a transistor is created in an IC circuit. Other components are created by these same techniques. B—A photoplotter creates integrated circuit artwork faster than by hand.
(Gerber Scientific, Inc.)

7. Using masking again we will control the etching away of the N-type island to create a new region, Fig. 4-18.

8. The wafer is exposed to another P-type diffusant and an area is created for the transistor's emitter region, Fig. 4-19. Resistors, diodes, and capacitors may also be created between these regions.

9. After the circuit is completed, a thin coating of aluminum is vacuum-deposited over the entire circuit. The aluminum is then etched to form patterns between the resistors, diodes, and transistors. The aluminum will also create pads for securing wires going to the external connections.

10. The wafer is then cut into separate circuits. This is a very simplified look at IC fabrication. There are also other methods and techniques for IC manufacturing. Scientists are now working on a chip created from grown protein. Advances are happening daily.

Advantages of IC circuits are their size, weight, cost, and reliability. The size of an IC is an advantage over the equivalent number of individual components. Size gives it a tremendous weight advantage. The cost of complete IC circuits are very often comparable to that of individual transistors. The IC has great reliability. It is 100 times more reliable than a single transistor. With all these advantages there are still some disadvantages.

Disadvantages are: It is difficult to create coils and capacitors in the IC package. They must work at low operating voltages and current ratings. The miniature diodes and transistors are delicate and cannot tolerate rough handling or excessive heat. The disadvantages are minor and insignificant compared to the advantages.

Some application for IC circuits are digital watches, pocket calculators, electronic games, stereo equipment, computers, and many other devices. The size and cost make IC circuits desirable for these applications.

CAPACITORS AND AC/DC COMPONENTS

A capacitor is a device consisting essentially of two conducting surfaces separated by an insulating material. The insulating material can be paper, mica, glass, plastic films, oil, or air. A capacitor stores energy, blocks the flow of direct current, and allows the flow of alternating current.

GENERAL CAPACITOR

Like the general resistor, the general capacitor has one fixed and set value. This value is established by the spacing, Fig. 4-20, and/or the size of the plates.

VARIABLE CAPACITOR

Variable capacitors can be adjusted by changing the useful area of the plates or the distance between them, Fig. 4-21.

POLARIZED CAPACITOR

Polarized capacitors can be put in the circuit in only one direction. The symbol should be placed to the plus polarity. The plus side will be the straight side of the symbol, Fig. 4-22.

Information for the capacitor should be written as shown in Fig. 4-23.

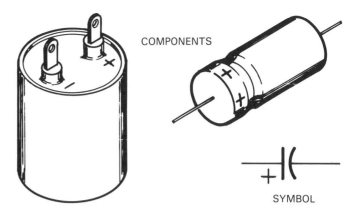

Fig. 4-22. A polarized (electrolytic) capacitor with its symbol. The plus end is indicated on the physical component. In order to buy a general capacitor you must tell the vendor three things: the value in farads, the voltage rating, and the tolerance.

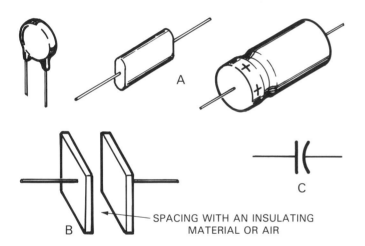

Fig. 4-20. A—Three of the many styles for common capacitors. B—The basic structure of a capacitor. C—A general capacitor symbol. Note now the symbol depicts the basic function.

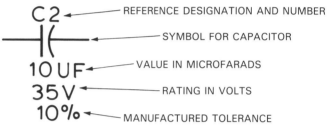

Fig. 4-23. A capacitor symbol with complete information.

COIL, CHOKE, OR INDUCTOR

The coil, choke, or inductor is a device made up of a coil of insulated wire around an iron, ceramic, or air core. See Fig. 4-24. It resists the changing of alternating current and its passage, but gives little opposition to the flow of direct current.

Coils are valued in henries, a unit of inductance. Resistance in ohms, and current-carrying capacity in amperes may also be listed, Fig. 4-25.

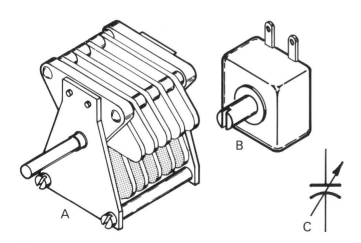

Fig. 4-21. A, B—Two types of variable capacitors. C—The symbol for a variable capacitor. Note the arrow for variable.

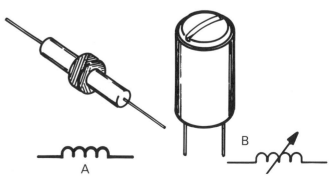

Fig. 4-24. A—A general coil and symbol. B—A variable coil and symbol.

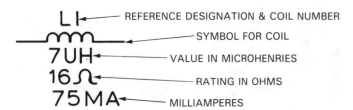

Fig. 4-25. A coil symbol with information.

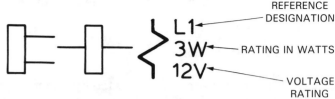

Fig. 4-27. Symbols commonly used for the solenoid.

SOLENOID

A solenoid is an electromagnetic device having an energizable coil and a magnetic core, Fig. 4-26. This core will move when the coil is energized. It performs mechanical functions. On our cars, it is used to engage the starter bendix gear when it is energized by turning the key to start the car.

Solenoids can be shown symbolically three ways, Fig. 4-27.

RELAY

A relay is an electromechanical device used to open and/or close contacts or switches as they are sometimes called. See Fig. 4-28. The part to operate the contacts is an electromagnet. It is a coil of wire around a soft iron core. The electromagnet moves a lever that opens or closes the contacts. Relays are used to start and stop many mechanical devices.

Relay symbols are shown differently from one company to another. They all describe the same device with some symbol variations, Fig. 4-29.

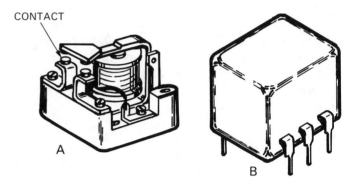

Fig. 4-28. A—An open relay showing contacts. B—An encapsulated relay used on printed circuit boards.

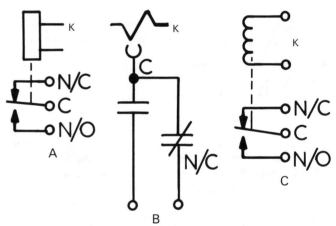

C = COMMON
N/C = NORMALLY CLOSED CONTACT
N/O = NORMALLY OPEN CONTACT

Fig. 4-29. The different ways to show a relay coil and contacts.

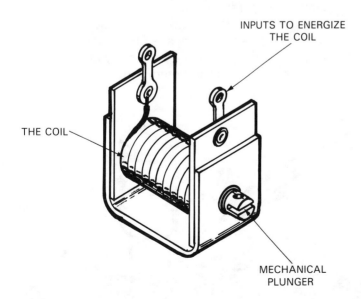

Fig. 4-26. A general solenoid. Solenoids use the same reference letter as the coil: "L."

TRANSFORMER

The transformer is another electromagnetic device, Fig. 4-30. By induction, it changes primary voltage and current values to different values on the secondary. The frequency remains the same.

A transformer has two coils or a tapped coil. One coil will be the primary section, the other the secondary. They can step voltage up or down.

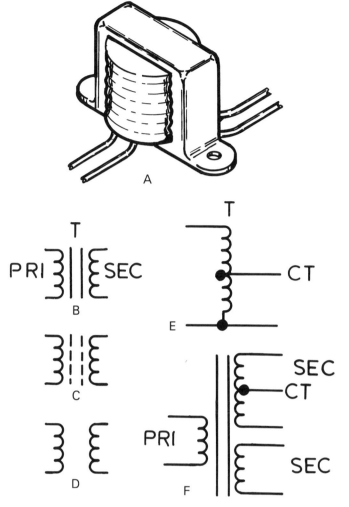

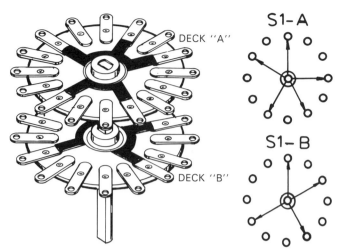

Fig. 4-31. A rotary switch with two decks. Each deck has multiple wipers which are ganged or mechanically coupled to the rotating shaft.

Fig. 4-30. A—A typical transformer. B—An iron core transformer symbol. C—A ceramic core symbol. D—An air core symbol. E—An auto transformer (single winding with a tap). F—A two secondary transformer with one being center tapped.

The transformers we see on the utility poles in older neighborhoods are the stepdown type. They step the voltage down to a level we can use in our homes. Most transformers used in electronics are also the stepdown type. They take the incoming 120 volts down to levels used by the electronics equipment.

SWITCH

The switch a is mechanical or electrical device that opens or closes a circuit. Switching may also be called making or breaking the circuit. There are many different types of switches. Fig. 4-31 shows a rotary switch. Other switch types are toggle, sliding, rocker, and precision, Fig. 4-32.

The closing of a switch is called making the circuit. The opening of a switch is called breaking the

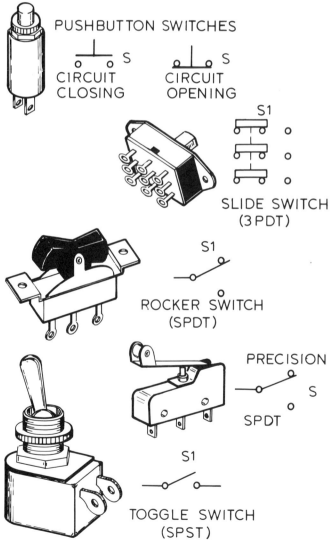

Fig. 4-32. The above switches show the major types used in industry and their symbols.

circuit. Terms such as single pole, double throw, break before make, are used in switching. Fig. 4-33 shows some of these symbol forms.

Switches are referenced with the letter "s." In order to purchase a switch we must indicate switch type, voltage, and amps. The switch information is presented in Fig. 4-34. The switch symbol should be drawn with the switch in its normal position. In the example, Fig. 4-34, the switch is a normally open type.

BATTERY

A battery is a direct current source made up of one or more cells. Refer to Fig. 4-35. These cells will convert chemical energy into electrical energy. The batteries contain the power supply for much of our portable electronics equipment. Calculators, transistor radios, and flashlights are some of the battery-operated devices you have used. Batteries are rated in volts and amps.

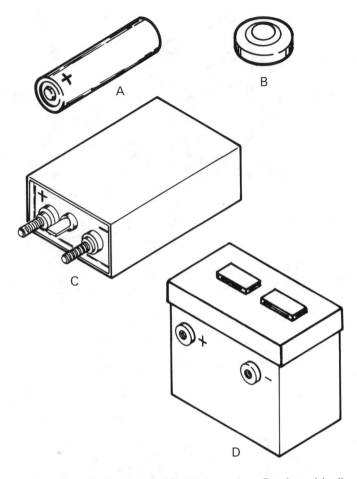

Fig. 4-35. A, B, C—Single cell batteries. D—A multicell battery.

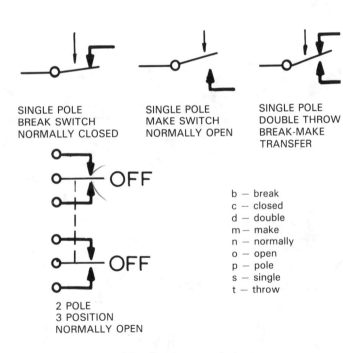

SINGLE POLE
BREAK SWITCH
NORMALLY CLOSED

SINGLE POLE
MAKE SWITCH
NORMALLY OPEN

SINGLE POLE
DOUBLE THROW
BREAK-MAKE
TRANSFER

OFF

OFF

b — break
c — closed
d — double
m — make
n — normally
o — open
p — pole
s — single
t — throw

2 POLE
3 POSITION
NORMALLY OPEN

Fig. 4-33. Common switch terms.

Battery symbols are completed with the information shown in Fig. 4-36. The long line on the symbol indicates the positive side but the " + " will normally be added for further clarification.

ANTENNA

Antennas may also be referred to as aerials. Antennas are used to receive or transmit radiating waves. There are various types of antennas, so you will use different symbols to indicate the use of each, Fig. 4-37.

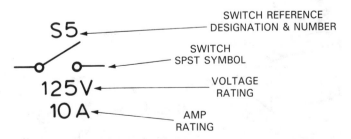

S5 — SWITCH REFERENCE
DESIGNATION & NUMBER

SWITCH
SPST SYMBOL

125V — VOLTAGE
RATING

10 A — AMP
RATING

Fig. 4-34. A switch symbol with necessary information.

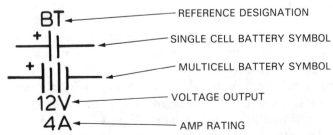

BT — REFERENCE DESIGNATION

SINGLE CELL BATTERY SYMBOL

MULTICELL BATTERY SYMBOL

12V — VOLTAGE OUTPUT

4A — AMP RATING

Fig. 4-36. Battery symbol with reference information.

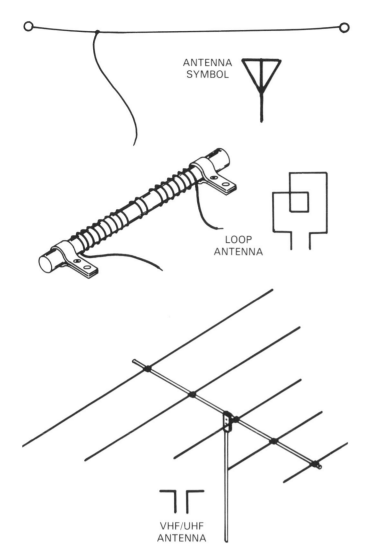

Fig. 4-37. Antenna types and associated symbols.

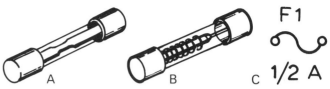

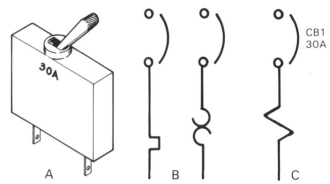

Fig. 4-38. A—General type fuse. B—Slow blow fuse. C—A fuse symbol representing a 1/2 amp fuse.

Fig. 4-39. A—Typical manually-operated circuit breaker. B—Thermal-overload symbol for circuit breaker. C—Magnetic-overload symbol with reference designation and amp rating.

FUSE

Protective devices are used to safeguard electronics equipment. Some of these are fuses. A fuse usually consists of a short piece of wire or metal which separates when the current exceeds its preset limits, Fig. 4-38. Fuses are rated in amps. Enough current causes the heat in the circuit that will burn out or melt the fuse wire. People commonly call this a blown fuse. If it were not for the fuses in a circuit, electronics equipment would be damaged and cause a far greater repair cost than replacing a fuse.

CIRCUIT BREAKER

A circuit breaker is another component used to protect electrical equipment, Fig. 4-39. Unlike the fuse, a circuit breaker will open an overloaded circuit without damaging itself. The circuit heat will cause it to open. Then as soon as the temperature is back to a normal operating range the circuit may be reclosed. Circuit breakers protect our homes. Most circuit breakers work by being thermally overloaded, but some use magnetic overload.

CRYSTAL

A crystal is a thin slab of quartz, Fig. 4-40. It is built with a preset thickness so it will vibrate at a specific frequency when energized. It is used as

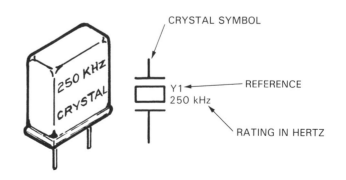

Fig. 4-40. A crystal and symbol with designation. This is a 250 kilohertz crystal. Hertz (Hz) means frequency or cycles per second. This crystal cycles 250,000 times per second.

a frequency control element in radio frequency oscillators. The channels of a citizens band radio are controlled by crystals.

OSCILLATOR

Oscillators generate alternating current. In radio frequencies, the alternating current may range from thousands to millions of cycles per second. An oscillator is the starting point for radio transmission. One style of oscillator is shown in Fig. 4-41.

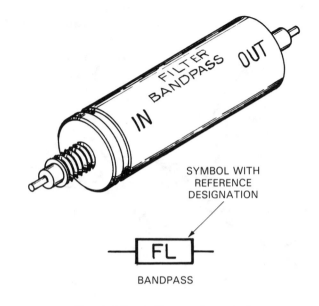

Fig. 4-42. A filter and symbol.

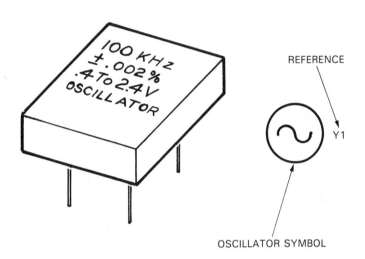

Fig. 4-41. An oscillator and symbol.

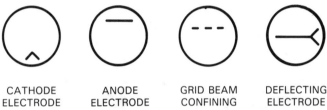

Fig. 4-43. Electron tube parts in symbolic representation.

FILTER

A filter is a component designed to separate wanted signals from unwanted signals or frequencies. The filters are used to suppress certain bands of frequencies while passing others easily. Three categories of filters are: high-pass, low-pass, and band-pass. High-pass will allow only high frequency passage. Low-pass will allow low frequencies passage. Band-pass will allow a range of frequencies, cutting out those in the high and low ends.

Filters come in many body types. See one body type in Fig. 4-42.

TUBE

Although tubes are being replaced by semiconductors, some are still in use. Tubes control electron flow much the way diodes and transistors do. They can amplify as transistors do and rectify as a diode does. Fig. 4-43 shows elements of tube symbols. Using these elements, you can create complete device symbols, Fig. 4-44. Tubes are much larger than their semiconductor counterparts.

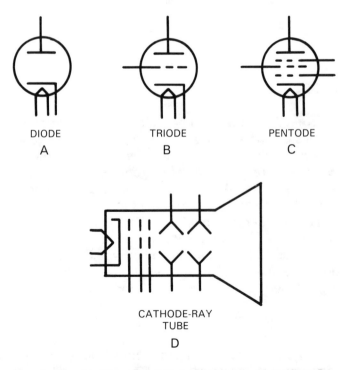

Fig. 4-44. A—The simplest type tube is a rectifier. B—Triode with heated cathode. C—A five element tube with three grids. D—A cathode-ray tube symbolically shown.

They create more heat during operation. This temperature requires a larger component so the heat can be dissipated. Most tubes are connected into the circuit by plugging into tube sockets, Fig. 4-45. This allows them to be replaced and checked easily.

CONNECTOR

A connector is any device on the end of a wire or cable to allow equipment to be connected to or disconnected from other equipment.

There are many types of connectors, but we use only a few symbols. See Fig. 4-46.

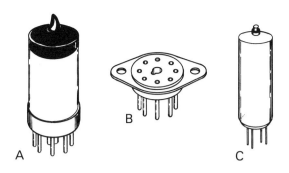

Fig. 4-45. A—A telephone tube. B—A keyed socket. Note: The center guide pin will allow the symmetrical connection to fit in only one position. C—Rectifier.

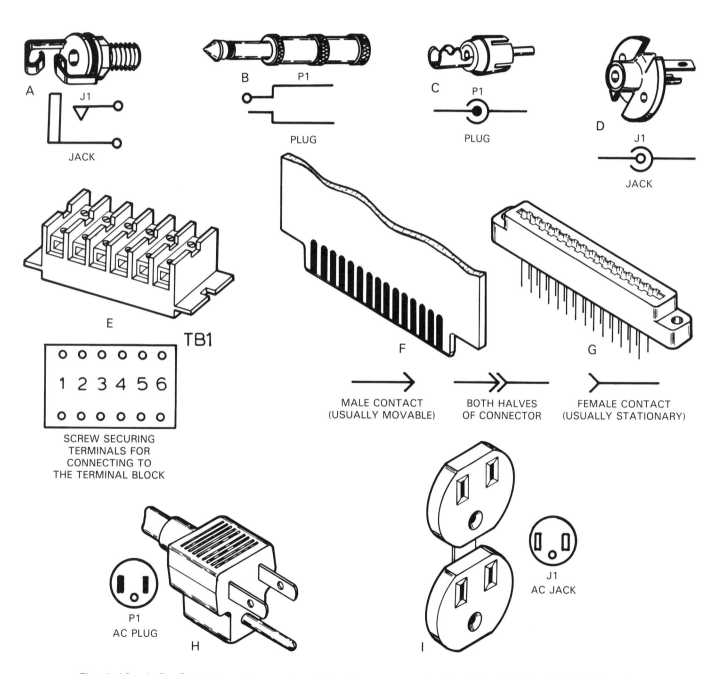

Fig. 4-46. A, B—Switchboard connector. C, D—Phono connector. E—A terminal block. F, G—Printed circuit board connector. H, I—Power supply connectors.

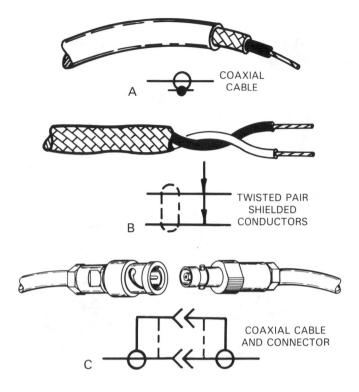

Fig. 4-47. A—A coaxial cable with symbol. B—A twisted pair with a shield. C—Coaxial plugs and cable.

CABLE, CONDUCTOR, OR WIRE

Cable may be referred to as conductor or wire. It comes in different styles for specific purposes. Types of cables and their symbols are shown in Fig. 4-47.

INPUT AND OUTPUT DEVICES

Electronics systems require an input and an output in order to complete a function. The inputs may be microphones or recording heads. Outputs may be speakers or headphones, Fig. 4-48. Each component is shown with symbol and reference designation.

A microphone is an electroacoustic tranducer, which responds to sound waves and delivers essentially equivalent electric waves to the amplifier. The speaker radiates acoustic power into the air with essentially the same waveform as that of the electrical input.

INDICATING, PILOT, AND SIGNALING LIGHTS

Lights perform different functions in electronics. They can be used as indicating lights. See Fig. 4-49. These lights will normally indicate such things as ''power is on,'' ''temperaure is too high,'' or there is some information needing to be indicated.

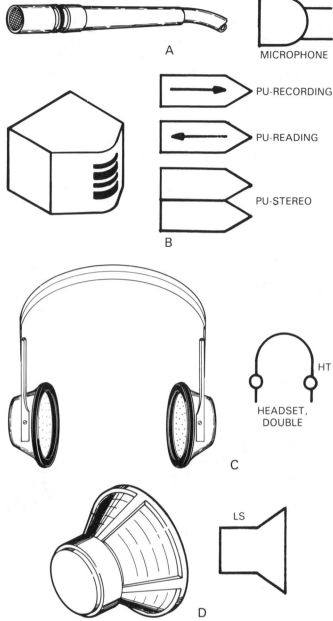

Fig. 4-48. A—A general microphone. B—Read, record, and stereo magnetic tape heads. C—Headphones. D—A speaker or loudspeaker. Each component is shown with symbol and reference designation.

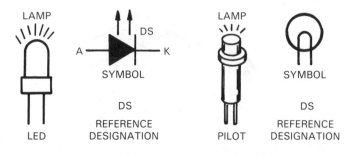

Fig. 4-49. Indicating lamps and accompanying symbols. Note LED lamp.

ILLUMINATING LIGHT

Area lights are lights which are used to light our homes and yards, Fig. 4-50. Lights which light up control panels so that meters and gages can be read are called illuminating lights. They are the same as area lights but normally smaller in wattage.

METER

Meters are used to show levels of current, frequency, speed, temperature, time, and other information. Examples of meters and their symbols are shown in Fig. 4-51.

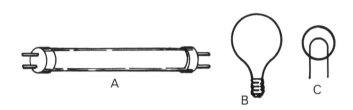

Fig. 4-50. Typical lamps. A—A fluorescent. B—An incandescent. C—The appropriate symbol. ''DS'' is the reference letter.

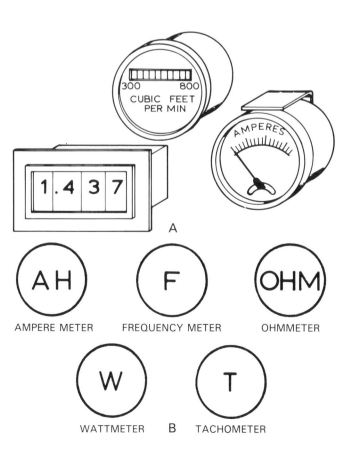

Fig. 4-51. A—Three types of meter faces. B—Symbols for standard meters.

ROTATING MACHINERY

Many of our electronics drawings will involve motors, generators, and their controlling circuits.

MOTOR

A motor is a machine that converts electrical energy into mechanical energy. It normally creates rotating power by turning a power shaft. Motors are used to drive sound equipment: phonographs, magnetic tape players, cooling fans, and many other applications, Fig. 4-52.

GENERATOR

A generator is a rotating machine which converts mechanical energy into electrical energy, Fig. 4-53. It can be used also to convert direct current voltage into alternating current of the desired frequency and amplitude.

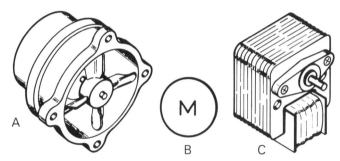

Fig. 4-52. A—Electric motor. B—Electric motor symbol and reference letter. C—Motor which can work as a combination motor-generator.

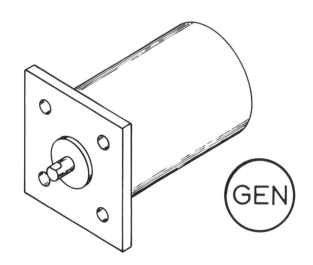

Fig. 4-53. A generator and symbol with reference designation.

CIRCUIT RETURNS

There are three symbols used for circuit returns. They are earth ground, chassis ground, and common ground symbols. Earth ground, Fig. 4-54A, is used to return the circuit directly to earth. AC circuits will use the earth ground symbol. Chassis grounds, Fig. 4-54B, are used to indicate circuits which return to the equipment's frame or chassis. The auto is a good example of a chassis ground unit. Common ground, Fig. 4-54C and D are used to show returns which have the same potential. This potential does not have to be zero. A common ground is sometimes called an airline.

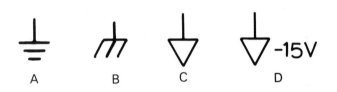

A   B   C   D

Fig. 4-54. A—Earth ground symbol. B—Chassis ground symbol. C—Common ground symbol. D—Common ground symbol with a modifier, which will make it common to other −15V sources of the drawing.

COMPONENT VALUES

There are preferred ways to write quantities in units such as ohms, volts, or henries. The values should be short and readable. Component values are expressed as shown in Fig. 4-55.

STANDARDS

All of the symbols and reference designations in this chapter comply with a standard. The two major standards are:
USAS Y32.16 Reference Designations for Electrical and Electronic parts and equipment.
USAS Y32.2 Graphic Symbols for Electronics and Electrical Diagrams.
Military standards are considered when military or government contracts are involved.

SYMBOL MODIFIERS

There are many things we can do to a basic symbol to change its meaning. Modifiers are used to change a component's meaning. You have seen some of the modifiers used earlier in this chapter. Note some additional modifiers and their uses in Fig. 4-56.

RANGE	VALUE	EXAMPLES
less than 1,000	ohm	.052
1,000 to 99,999	ohms or kilohms	1800 written 1.8 K 45000 written 45 K
100,000 to 999,999	kilohms or megaohms	100,000 written 100 K or 0.1 M
1,000,000 or more	megohms	10,000,000 written 10 M

A

	RANGE	VALUE	EXAMPLE
CAPACITORS	10,000 or more pico farads	MICRO farads	0.5 UF or 0.5 MFD 30 UF
	Up to 9999 pico farads	PICO farad	150 PF
INDUCTORS	.001 to .099 henrys	MILLI HENRYS	.005 H written 5 mH
	.1 or more henrys	HENRYS	15 H

B

Fig. 4-55. A—How to write resistor values. Symbol K will be hand-lettered in capitals. B—How to write capacitor and inductor values.

Adjustable components. We saw this modifier used with resistors, capacitors, and coils.

Radiation indicators. These modifiers are used to indicate the radiation of light, heat, or signals.

Examples: LDR (A light dependent resistor) and a light emitting diode.

Test point. A point in the circuit where we measure the current flow.

Polarity. Used to indicate which direction a device is installed in the circuit.

Arrow. The arrow is used to indicate the signal or power flow.

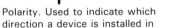

Fig. 4-56. Modifiers used to add meaning to basic symbols.

REVIEW QUESTIONS

1. What function does a resistor perform?
2. What regulates the amount of resistance?
3. What phrase reminds you of the NPN type of transistor?
4. Using the resistor color code, (appendix), state

the value for the following resistors.
   a. brown black brown silver
   b. orange green orange gold
   c. brown green orange silver
   d. orange black green gold
5. Provide the colors for the following:
   a. 270 ± 5%
   b. 2400 ± 10%
   c. 4.7K ± 10%
   d. 5.6K ± 5%
   e. .18M ± 5%
   f. 1.1M ± 5%
6. Explain how a rheostat works.
7. A capacitor blocks _____ (AC, DC).
8. What information must be given when purchasing a capacitor?
9. What does a coil do?
10. How many symbols are used to show solenoids?
11. What are the two sections of the transformer?
12. Relays perform what functions?
13. What does - - - - - - - -mean when placed between two adjustable symbols?
14. What type of current source does a battery provide?
15. What is the main difference between a fuse and a circuit breaker?
16. What are two ends of a diode?
17. How are zener diodes used?
18. Tubes have been replaced by what components?
19. What does it mean to you when it is stated— a connector is Keyed?

20. What does it mean to correctly reference a resistor? List about three ideas.

PROBLEMS

PROB. 4-1. Draw a resistor symbol and provide all identifying information.
PROB. 4-2. Practice drawing a transformer symbol. Add a center tap symbol. Provide all needed information.
PROB. 4-3. Using your symbol template, create the following components: Label each with the appropriate reference designation.
   1. Transistor (PNP).
   2. Loop antenna.
   3. Diode (Zener).
   4. Potentiometer used as rheostat.
   5. Transformer (iron core)
   6. Tapped resistor.
   7. Transistor unijunction.
   8. Fuse.
   9. Chassis ground.
   10. Coaxial cable.
   11. Battery multicell.
   12. Circuit Breaker.
   13. Inductor.
   14. Capacitor (Variable).
   15. Switch (mechanically coupled) (rotary).
   16. Speaker.
   17. Microphone.
   18. Pickup head.
   19. Motor.
   20. Transistor (NPN).

Chapter 5

SCHEMATIC AND LOGIC DIAGRAMS

After studying this chapter you will be able to:
■ Use symbols in schematic circuits.
■ List the qualities of a good schematic and draw a schematic.
■ Work from an engineering sketch.
■ Reference a component on a schematic.
■ Draw a logic diagram and interpret logic symbols.
■ Know the function of AND, OR, NAND, NOR, and INVERTER gates.

In this chapter you will be studying two similar drawings: schematic and logic diagrams. These drawings usually come after the preliminary drawings, the block and single line diagram.

SCHEMATIC DIAGRAMS

The schematic drawing is a symbolic representation of the data and components used in an electronic circuit. You will learn how the schematic uses symbols from the electronic language. Service, sales, manutacturing, and engineering people cannot adequately communicate about an electronics device without the help of a schematic or logic drawing. The way the components are connected will inform the reader of their function. The function will be further explained when you add values, ratings, tolerances, and catalog numbers to the components, Fig. 5-1.

Engineers and drafting supervisors give you sketches of the schematic from which to work. The sketches given to you as a beginning drafter will be formally laid out. Then as your skills and knowledge increase, the sketches will be only rough layouts of the circuit, Fig. 5-2. A schematic drawing can be laid out in many ways. No two drafters will organize the drawing exactly the same way. However, they should follow some basic principles. These principles will be discussed later in this chapter.

DRAFTING SCHEMATICS

When drafting a schematic, your responsibilities are:
1. To organize the schematic to fit a specified paper format. Often schematics must fit into books. This requires you to draft your drawings to a size that will fit the book or can be reduced appropriately to fit the book.
2. To apply proper symbols and reference data for each component. This will be accomplished by having an understanding of American National

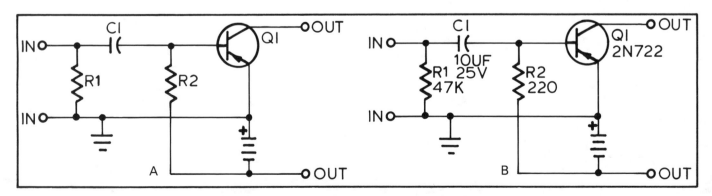

Fig. 5-1. A—Schematic of a transistor amplifier circuit. B—Add more information chiefly to help a technician troubleshoot the circuit.

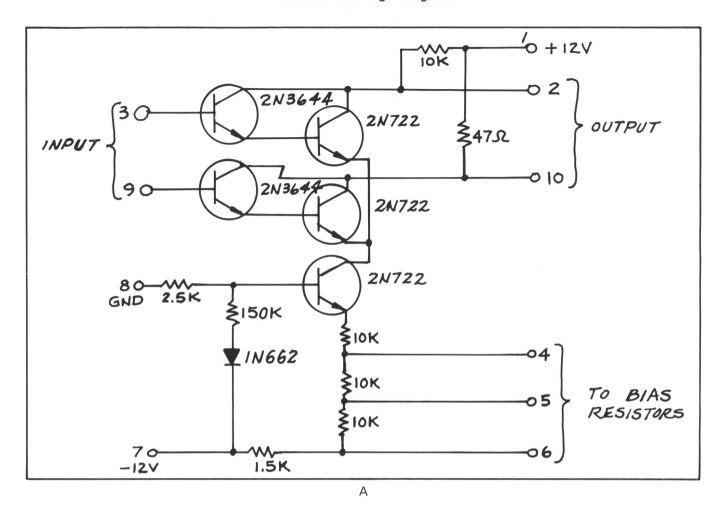

A

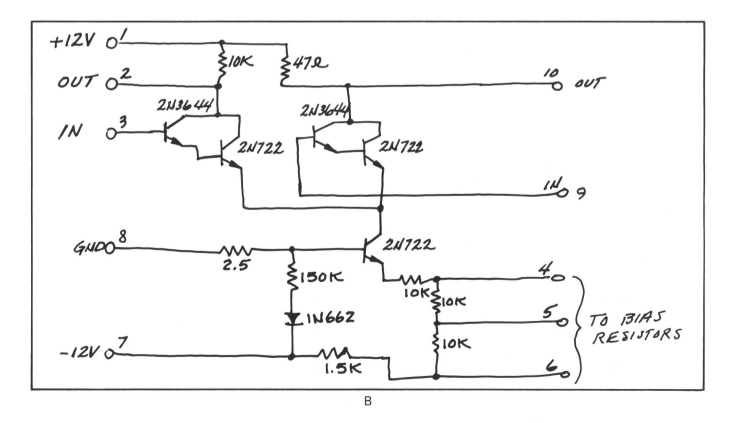

B

Fig. 5-2. A—Formal engineering input to the drafter. B—An engineering input submitted as a rough sketch.

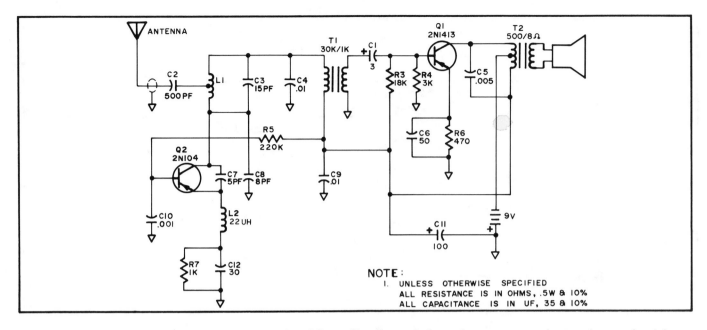

Fig. 5-3. A schematic showing left to right signal flow. Signal travels from the antenna to the speaker on the right.

Standard Institute's (ANSI) Y32.16 Reference Designations for Electrical and Electronics, Y32.2 Graphic Symbols for Electrical and Electronics Diagrams, and Y32.14 Logic Diagrams, Graphic Symbols (AIEE/IEEE91).

3. To create general notes, legends, or detailed notes that explain specifics of the schematic.
4. To follow the general rules when drafting the schematic.

RULES FOR DRAFTING SCHEMATICS

All drawings have general rules for the way they are created. The schematic should be drawn using the following rules:

1. Normal signal flow should be from left to right, top to bottom. An example would be a radio circuit, Fig. 5-3. The antenna (input) should be on the upper left of the paper. The speaker (output) should be on the right side.
2. Lines should be spaced a minimum of 3/8 in. (10 mm) apart.
3. Lettering should be 5/32 in. (4 mm) high. This requirement is necessary for microfilming and photo-reductions of the drawings.
4. Lines between components should take the shortest path.
5. Connecting lines should have a minimum of crossovers and joggles, Fig. 5-4 A and B.
6. Long parallel lines should be arranged in groups, preferably three to a group, Fig. 5-4 B and C.
7. Avoid four-way tie points or four-way junctions, Fig. 5-5.
8. Power sources should go up and ground lines

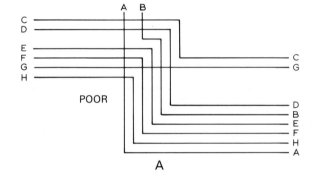

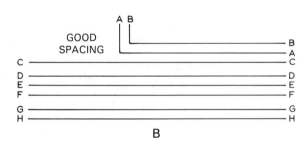

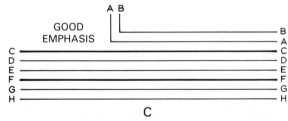

Fig. 5-4. An example of how to handle jogged, crossover, and parallel lines. A—The eye cannot easily follow a line. B—Regrouping and periodic spacing helps. C—Periodic thick lines are one common method of improving readability.

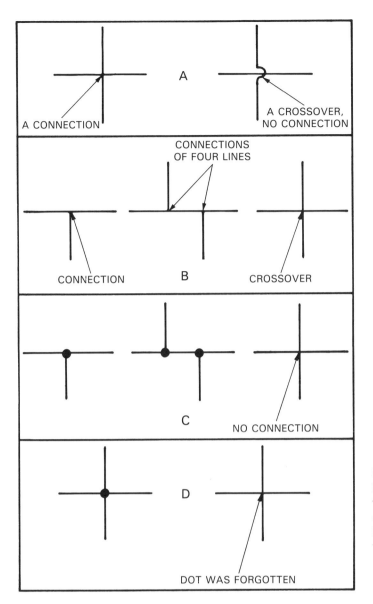

Fig. 5-5. Junctions and crossover methods used on schematics. A—An outdated method. B—Method without dot makes only one junction at a given place. C—Method using single junction with dot is most preferred. It helps trip reader's eye when it scans a line. D—Avoid four-way tie points in case you forget the dot.

should go down, as shown in Fig. 5-6.

9. All lines will run horizontally or vertically and connect in 90° corners, Fig. 5-7. A flip-flop or crossover circuit is the only exception to this rule.

REFERENCE DESIGNATIONS

Reference designations are combinations of letters and numbers, Fig. 5-8. They are used to identify components shown on the schematic. The reference designation should be located as close to the graphic symbol as possible.

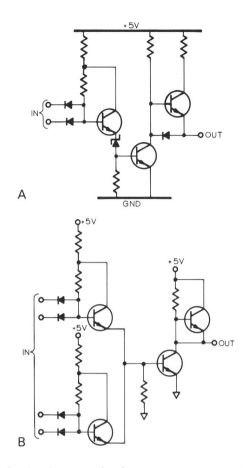

Fig. 5-6. A—An example of power sources up and ground lines down. Some engineers prefer the heavy bus lines for the ground and power ties. B—When many lines will be crossed, use a common ground symbol and power symbol. Note: The +5V of the bottom circuit does not go to the top of the paper. It simply goes up. This saves crossing lines and makes the drawing easier to read.

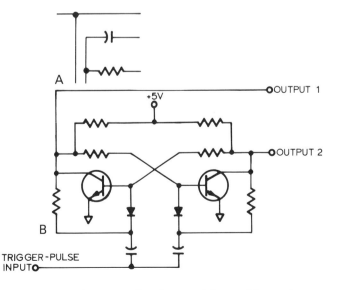

Fig. 5-7. A—An example of normal lines. All are drawn horizontally or vertically with connections made at right angles. B—A flip-flop circuit is an exception. It uses diagonal lines to show the crossover function.

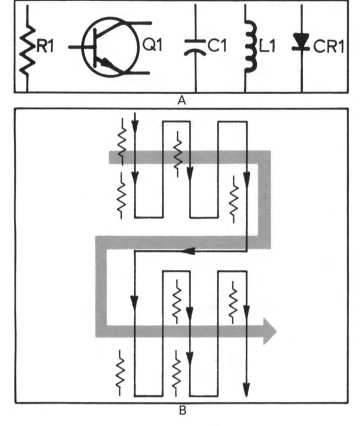

Fig. 5-8. A—An example of components and their reference designations. B—How to number the components in the proper order.

Numbering of components should be in a sequential order starting from upper left and proceeding left to right, top to bottom. When items are eliminated because of drawing revisions, the remaining items are not renumbered. But you create a table to show the missing or sometimes forgotten numbers, Fig. 5-9.

REFERENCE DESIGNATIONS	
LAST NUMBER USED	NUMBERS NOT USED
R86	R53
C29	
Q23	Q15, Q12
L6	
S1	

Fig. 5-9. A typical table showing the last reference number used and numbers not used.

Placement of references will be dictated by each company's drafting standard. Most companies will follow the methods shown in Fig. 5-10. To save room on a crowded schematic, some companies prefer the method shown in Fig. 5-11.

SERIES AND PARALLEL CIRCUITS

In order to understand how to rearrange and improve a schematic you must first understand parallel and series circuits. See Figs. 5-12 and 5-13. A parallel circuit will have one end of the components going to a common source and the other ends going to another common source. A series circuit will have the components connected end to end. A series circuit does not allow much flexibility in rearrangement but parallel circuits can be moved around more readily, Fig. 5-14. Schematics are normally made up of individual components. Our next drawing, logic circuits, will be made up of components integrated into single units called integrated circuits.

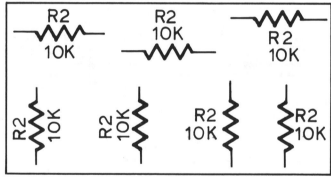

Fig. 5-10. Commonly used methods of presenting symbols and reference information.

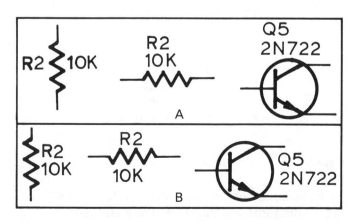

Fig. 5-11. When the schematic is crowded, and lines must be spaced a minimum distance apart, method (B) is the preferred way of referencing the components.

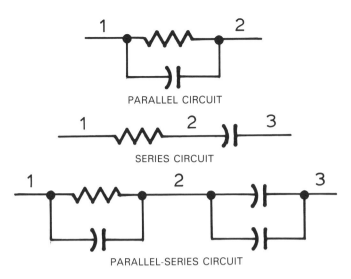

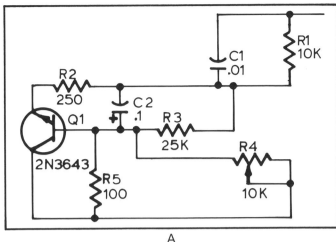

A

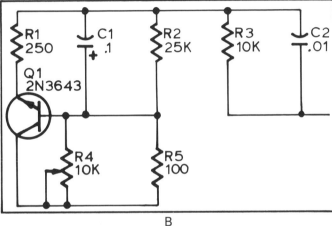

B

Fig. 5-14. A parallel-series circuit shown before, A, and after the drafter rearranged it. Example B lines the components up so the lettering can be done on a common line, making the circuit easier to read.

Fig. 5-12. Series and parallel circuits. The potential will change at each junction. Note the numbers on the examples. Each number represents a new or different potential.

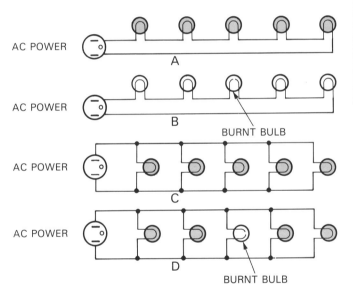

Fig. 5-13. Effect of series circuit and parallel circuit. A—Lamps in series. B—String of lamps are all out when in series circuit. C—Lamps in parallel. D—String of lamps are on except for burnt bulb in parallel circuit.

LOGIC DIAGRAMS

Logic drawings are diagrams representing the logical elements and their interconnections. They will most often not show the component's elements or internal details. They will use symbols and supplementary data to describe the function of each element. Symbols for logic diagrams are covered in Graphic Symbols for Logic Diagrams (two-state devices) ANSI Y32.14.

TYPES OF DIAGRAMS

There are two main types of logic diagrams, a basic and a detailed diagram, Fig. 5-15. The basic diagram shows logical functions and their relationships without reference to physical relationships. It uses logic symbols to show the main concept of the circuit. Detailed diagrams take the basic information and add specifics or nonlogic data. This data may include pin numbers, test points, and other necessary physical elements.

LOGIC ELEMENTS

There are a few elements used in logic diagrams. They are: AND, NAND, OR, NOR, and INVERTER GATES, plus Operational Amplifiers, Flip-Flops, Schmitt triggers, Decoders, Counters, Shift registers, and Oscillators. These are some of the most frequently used elements.

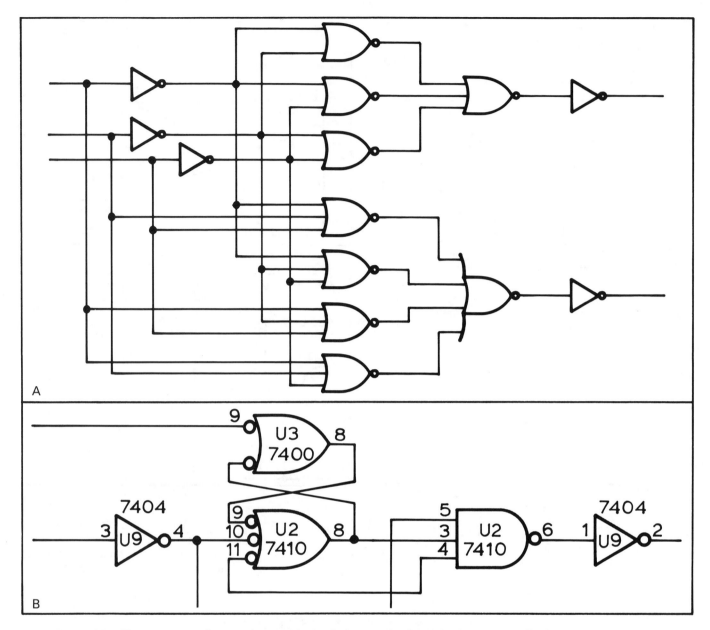

Fig. 5-15. The two most frequently used logic diagrams. A—Basic logic diagram. B—Detail logic diagram.

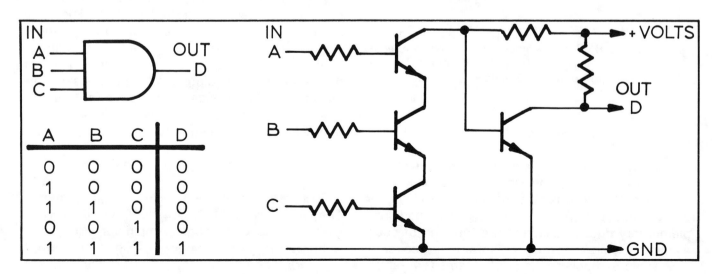

Fig. 5-16. An AND gate symbol, its schematic, and truth table.

LOGIC TERMS

AND gate—A gate circuit with more than one input terminal, Fig. 5-16. No output signal will be produced unless a pulse is applied to all inputs simultaneously. In binary circuits all inputs must be "1" to get an output "1". If any of the inputs are zero, the output will be zero.

INVERTER gate—A circuit that takes a positive signal input and puts out a negative signal, or vice versa, Fig. 5-17. It has one input and one output. It is often called a NOT circuit since it produces the reverse of the input.

NAND gates—A combination of a NOT and AND function, Fig. 5-18. It has two or more inputs and one output. The output is logic "0" if all the inputs are "1". If any input goes to "0" the output goes to "1". With logic of the opposite polarity, this type gate becomes a NOR gate.

OR gate—The OR circuit performs the function of producing an output whenever any one (or more) of its inputs is energized, Fig. 5-19.

NOR gate—A combination of an OR and NOT function, Fig. 5-20. It will have an output of "0" if any input is logic "1" and is logic "1" only if all the inputs are logic "0". With an opposite logic polarity, this gate can become a NAND gate.

Operational Amplifier (Op Amp)—An amplifier that performs various mathematical operations. They can be used to add, subtract, average, integrate, and differentiate. It may have a single input and output, Fig. 5-21.

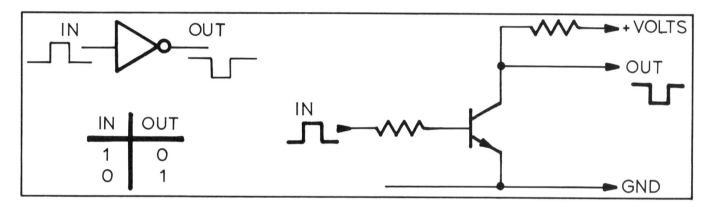

Fig. 5-17. An INVERTER (NOT) symbol, its schematic, and truth table.

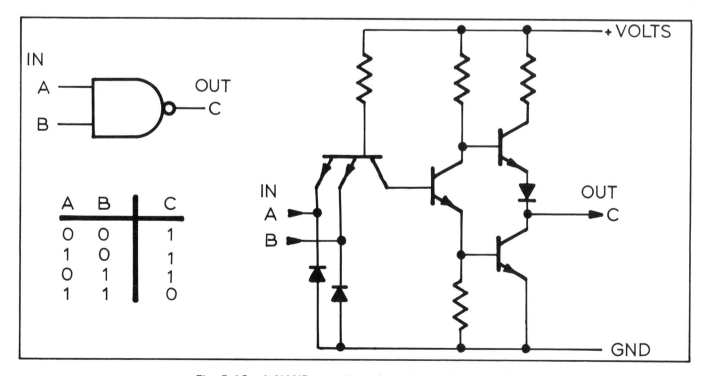

Fig. 5-18. A NAND gate, its schematic, and truth table.

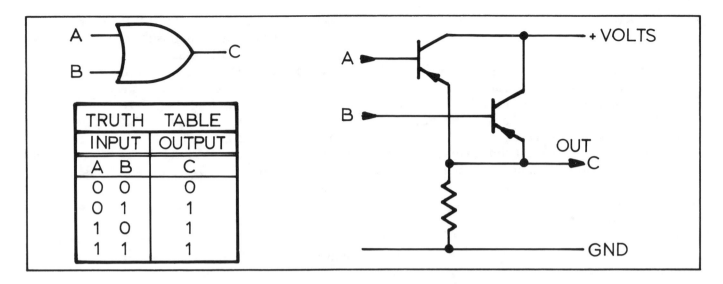

Fig. 5-19. An OR gate with schematic and truth table.

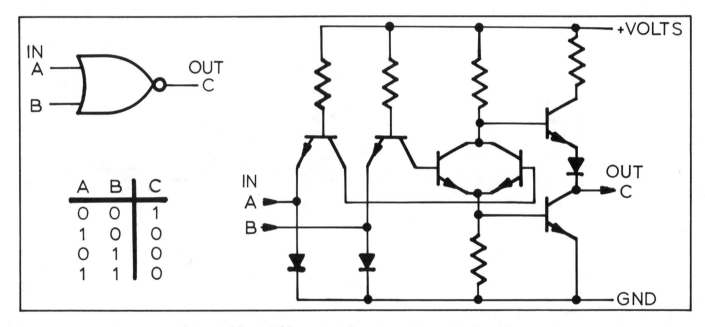

Fig. 5-20. A NOR symbol, its schematic, and truth table.

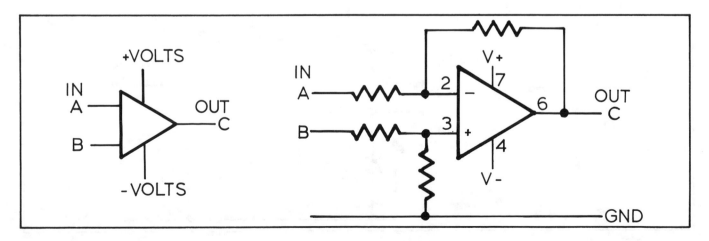

Fig. 5-21. A two input Operational Amplifier (Op Amp) with external resistors added to the symbol.

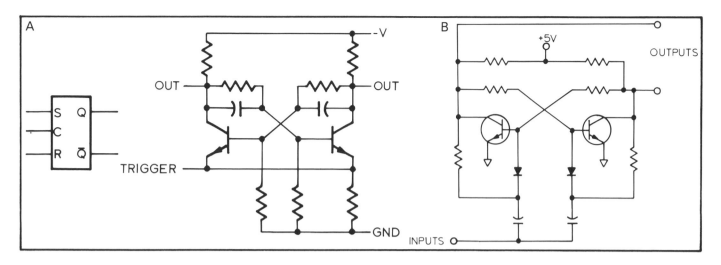

Fig. 5-22. A—Flip-flop symbol and a schematic of the flip-flop function. B—Another type of flip-flop.

Flip-Flop—A flip-flop, Fig. 5-22, is a device which is stable in either of two states. When triggered by an input or clock pulse, the flip-flop moves from one stable state to the other. For example, if one of its outputs, called Q, starts at 1, it will be at 0 after the input pulse.

The most common flip-flop is the JK type. Another type is the set-reset (RS) flip-flop. Some types require both an input pulse and an ''enable'' pulse.

Schmitt Trigger—A bistable pulse generator in which an output pulse of constant level exists only as long as the input is constant, Fig. 5-23.

Decoder—A device for translating a combination of signals into one signal. It is often used to extract information from a complex signal or coded signal, Fig. 5-24.

Counter—A logic device that counts input pulses, Fig. 5-25. It may count input pulses and then output after a predetermined number has been

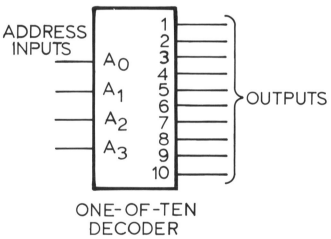

Fig. 5-24. A decoder with four inputs and ten outputs.

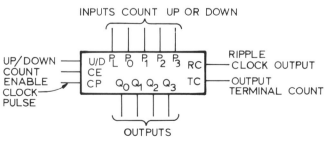

Fig. 5-25. A counter circuit symbol. The device can count up or down.

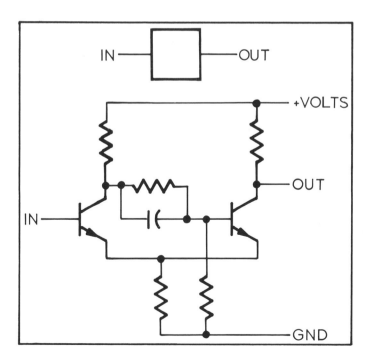

Fig. 5-23. A Schmitt trigger symbol with a schematic.

received. The counter may count up or down.

Shift register—A circuit which can shift information from one flip-flop in a chain to an adjacent flip-flop when it receives a clock pulse, Fig. 5-26.

Oscillator—An electronic device that generates alternating current of predetermined frequencies,

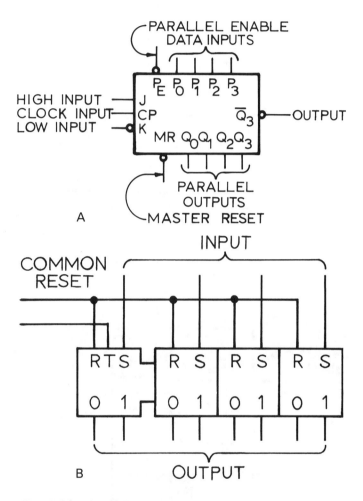

Fig. 5-26. A—Shift register symbol. The unit is used for data transfer. B—Four flip-flops used as a four-stage shift register.

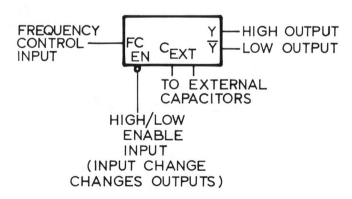

Fig. 5-27. The frequency control of an oscillator sets its frequency.

Fig. 5-27. The oscillator circuit is designed to take a voltage impulse and produce a current that periodically reverses. To properly draw oscillators and other elements correctly in the logic diagram, you must follow some basic rules.

Rules for drawing logic diagrams

1. Draw each device so that the input is on the left or top of the element.
2. Outputs of logic elements should go to the right or down.
3. The basic rules for schematics apply also to logic diagrams.
4. The numbering of logic elements will be by physical positions in the equipment. This rule will differ from the schematic which is numbered left to right and top to bottom disregarding physical position.

REVIEW QUESTIONS

1. Long parallel lines should be put in groups of _____(how many).
2. List the drafter's responsibility when drawing a schematic.
3. Which way should the schematic's signal flow?
4. Why should we avoid four-way tie points?
5. What are reference designations?
6. In what symbol can diagonal wiring lines be used?
   a. AND
   b. Op Amp.
   c. NOR
   d. Flip-flop.
7. What is a parallel circuit? Sketch one.
8. What is the standard source for symbols for logic diagrams?
9. What are the two main types of logic diagrams?
10. List the basic logic elements.

PROBLEMS

PROB. 5-1. Draw a formal schematic of the panel-mounted components for the AM-FM radio shown in Fig. 5-28. Draw on ''A'' size vellum. Add reference designations and make appropriate changes to the symbols.

PROB. 5-2. Prepare a schematic of the AM transmitter shown in Fig. 5-29. Sketch the circuit layout before going to the final drawing.

PROB. 5-3. Draw a schematic of the transistor amplifier shown in Fig. 5-30. Add correct reference designations for all components.

PROB. 5-4. Create a formal schematic of the push-pull amplifier, shown in Fig. 5-31.

PROB. 5-5. Draw a schematic of the oscillator cir-

cuit shown in Fig. 5-32.

PROB. 5-6. Create a logic drawing of the partial circuit shown in Fig. 5-33.

PROB. 5-7. On an "A" size sheet, make a drawing of the logic circuit shown in Fig. 5-34.

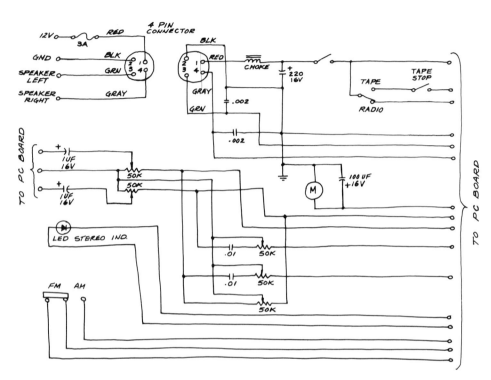

Fig. 5-28. Draw a schematic of an AM-FM radio's panel-mounted components.

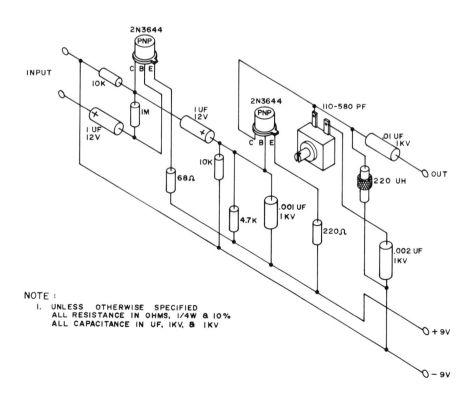

Fig. 5-29. Draw a schematic of an AM transmitter circuit.

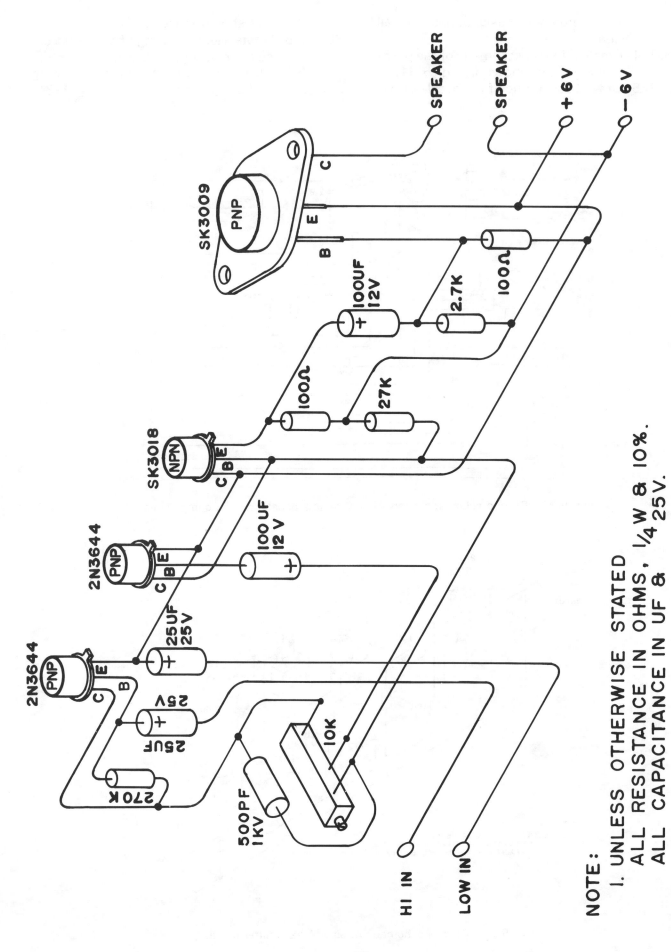

Fig. 5-30. Draw a schematic of transistor amplifier.

NOTE:

1. UNLESS OTHERWISE STATED
   ALL RESISTANCE IN OHMS, 1/4 W & 10%.
   ALL CAPACITANCE IN UF & 25V.

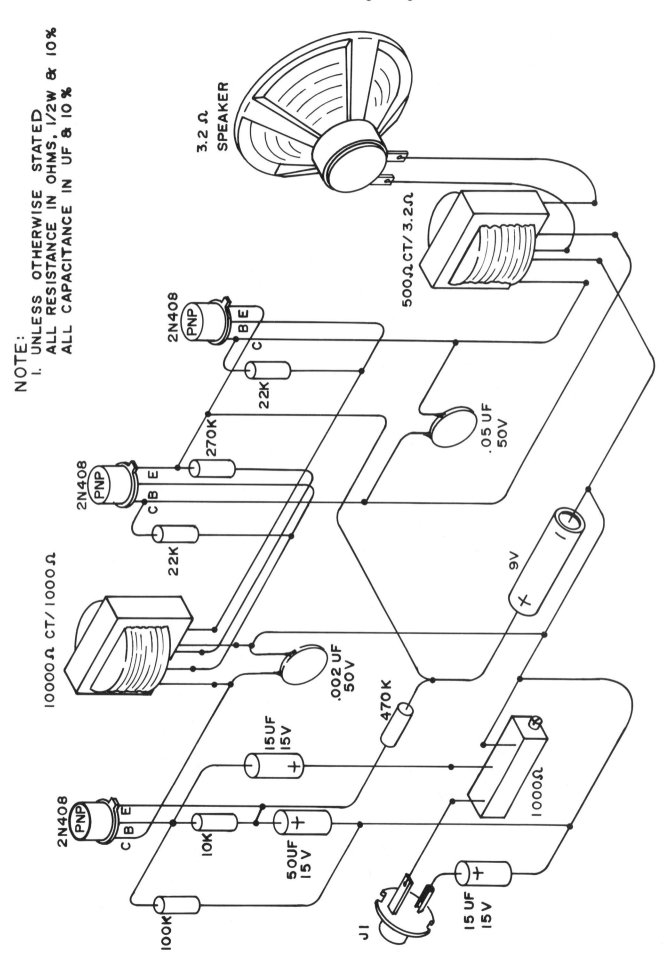

NOTE:
1. UNLESS OTHERWISE STATED
   ALL RESISTANCE IN OHMS, 1/2W & 10%
   ALL CAPACITANCE IN UF & 10%

Fig. 5-31. Draw a push-pull amplifier schematic.

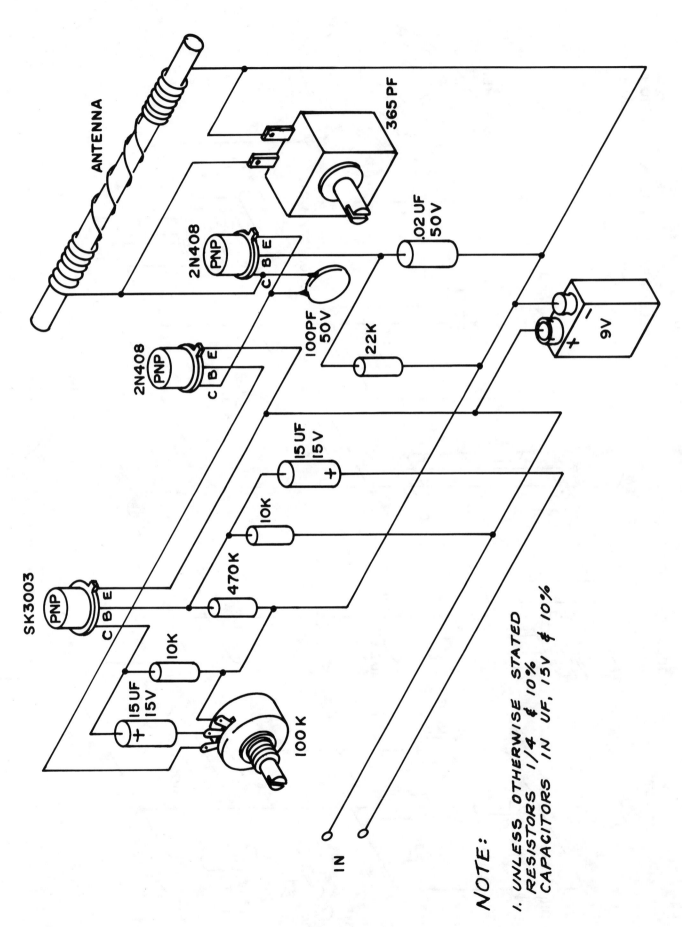

Fig. 5-32. Draw an oscillator circuit.

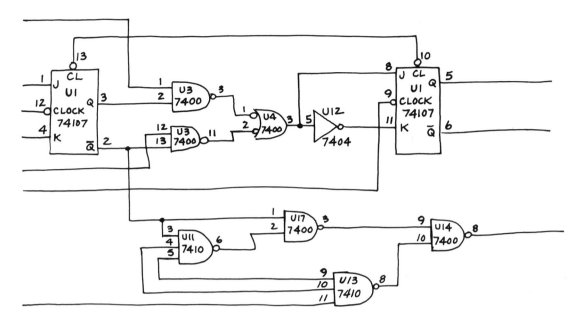

Fig. 5-33. Redraw a portion of a logic circuit.

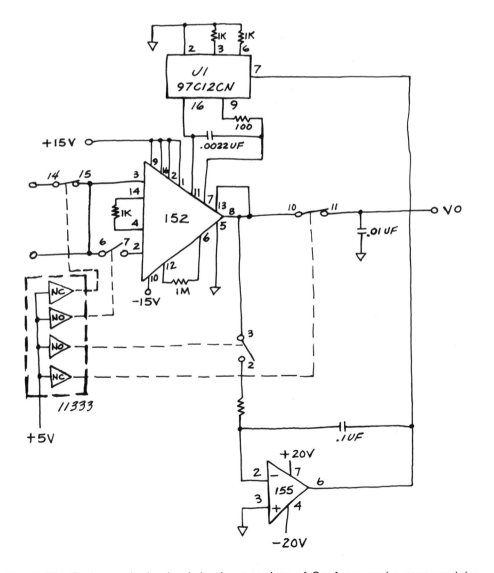

Fig. 5-34. Redraw a logic circuit having a variety of Op Amps and output modules.

Chapter 6

WIRING DIAGRAMS

After studying this chapter you will be able to:
- Select wiring methods.
- Create a wire list.
- Draft a Point-to-Point diagram.
- Draft a pictorial Point-to-Point diagram.
- Draw a highway diagram.
- Draw an interconnection diagram.
- Draw a cable assembly.
- Make harness assembly drawings.
- Select wire termination methods.

The majority of electronic equipment needs some kind of wiring interconnections. Understanding how to document this wiring is important knowledge for the drafter. This chapter will cover the basic methods used by the major companies.

WIRING METHODS

There are many ways to show factory workers, service personnel, and others in the engineering family how to wire electronics equipment. Normally the decision to choose one method over another is based on three different things:
1. Knowledge of the technician.
2. Quantity to be built.
3. Complexity of the equipment to be manufactured. Here are some of the ways drafters document wiring.

SCHEMATIC

The simplest method to get a wiring project built is to give a top qualified technician a schematic. The schematic will only supply the "from" and "to" information. This information will describe where the wire will be hooked (the from position) and where it is going (the to position), Fig. 6-1.

From a schematic, a drafter can create a wire list or other wiring documents. The wiring list will save the technician time in reading the schematic.

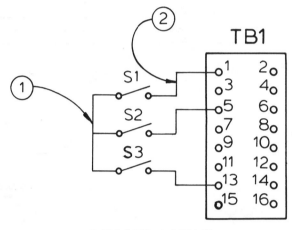

SCHEMATIC

Fig. 6-1. A schematic being used to wire between components. Note 1: Technician runs a wire connecting all switches. Note 2: Wire is run from S1 to TB1-1 (the terminal block).

WIRE LIST

The wire list is another elementary document. It will include the information from the schematic plus some additional information such as:
1. Color of wire.
2. Gage of wire.
3. Length of wire.
4. Entry in parts list.
5. Condition at terminals.

Fig. 6-2 shows how the wire list information is presented. To understand the wire list information, you need to be familiar with wires and their terminations.

WIRES OR CONDUCTORS

Conductors used in wiring electronics equipment are of three different types: solid, stranded, and flat

WIRE LIST

WIRE NO.	SIZE AWG	ITEM NO.	COLOR	FROM	CONDITION	TO	CONDITION	LENGTH
1	22	1	R	XA1-22	SOLDER	EOT-2	#6 LUG	20
2	22	2	W	XA1-21		P1-6	#6 LUG	14
3	20	3	O	XA1-19		P2-20	#6 LUG	16
4	20	4	Y	XA1-18		S3-8	SOLDER	14
5	18	5	BK	XQ1-E		S3-7	SOLDER	12
6	16	10	BR	XQ1-B		TB1-1	SOLDER	8
7	18	6	Y	XQ1-C		TB1-2	SOLDER	10

①   ②   ③   ④   ⑤   ⑥   ⑦   ⑧   ⑨

1. EACH WIRE IS IDENTIFIED BY A NUMBER.
2. SIZE OF WIRE BY AMERICAN WIRE GAGE.
3. ITEM NO. IN PARTS LIST (SEE BELOW)
4. COLOR OF WIRE USING STANDARD ABBREVIATIONS
5. COMPONENT & TERMINAL WIRE IS COMING FROM.
6. TYPE OF CONNECTION BEING MADE
7. COMPONENT & TERMINAL WIRE IS GOING TO
8. TYPE OF CONNECTION BEING MADE
9. LENGTH OF EACH WIRE (IN INCHES)

PARTS LIST

ITEM NO.	NOMENCLATURE	QTY
1	WIRE #22 AWG RED TEFLON COATED STRANDED.	36"
2	WIRE #22 AWG WHITE	20"
3	WIRE #20 AWG ORANGE	32"
4	WIRE #20 AWG YELLOW	

Fig. 6-2. Wire list and accompanying parts list. Note: Parts list numbers are used in column 3 of the wire list.

ribbon, Fig. 6-3. Solid conductors have traditionally been used where they will not experience bending. They are less expensive than stranded wires, but have a more limited use. In electronics, solid wiring is used mainly for jumpers (bus wiring) and for a process of wire wrapping.

Stranded wire has superior handling and flexing qualities. This makes it the most universally used wire. Refer to Fig. 6-4. The endurance of the stranded wire is judged by the number of strands

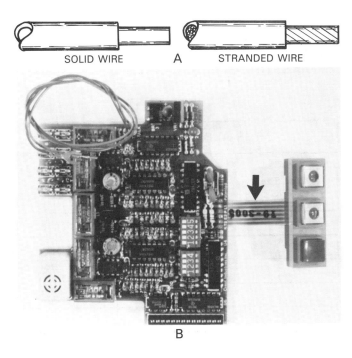

Fig. 6-3. A—An example of stranded and solid conductors. B—Stranded round wire and flat ribbon cable. (See arrow.)

Fig. 6-4. A cable termination on a circuit board saves space on the board edge. A flexible mounting and stranded wire help reduce stress on board and on connector. (Amphenol Products)

71

it contains. The larger the number of strands, the greater its bending endurance. Thinner wire will bend better than larger diameter wire.

The diameter of a wire will decide its gage. The wire's gage is determined by the American Wire Gage Standard, Fig. 6-5. Stranded wire will be identified by two numbers. The first number states the number of strands in the wire. The second number states the gage of each strand. Example 7/26 means 7 strands of #26 AWG wire. Number 26 gage wire is .0159 inches in diameter, Fig. 6-5.

The wire gages go from #4/0—the largest, to #44, the smallest. The table shown in Fig. 6-5 has only given the even sizes. It also has been condensed to show the most used wire sizes.

Wire length and diameter of wires affect both the resistance and current-carrying ability. The smaller diameter wires have more resistance to electron flow, and therefore less capability to carry current loads, Fig. 6-6.

BUS WIRE

Bus wire is bare wire (without insulation) normally used to make short terminal-to-terminal connections. It is solid wire, so it will be used where bending does not take place after installation. Where the bus wire requires insulation, a tube type insulation is slid over it. This tube insulation is called SPAGHETTI. The reason for using spaghetti is to avoid having to strip both ends of a short wire.

SHIELDED AND COAXIAL CABLES

Shielded or coaxial wires are used to exclude or contain undesirable radiation, Fig. 6-7, A and B. An example for the use of a coaxial wire is in an automobile radio antenna system. The shield around the signal wire keeps the unwanted engine and electrical noise from radiating into the signal wire. Without this shield, we would hear many interfering noises. The shield of the coaxial cable is grounded so the interfering electrical energy or radiation will be absorbed in the chassis where the shield is grounded.

WIRE TERMINATION

In order to make a wire useful, we must be able to electrically secure it where we desire. There are three basic ways to secure or terminate wires: Soldering, crimping, and wrapping. The method used is dictated by the terminal to which the wire is to be secured. You can show the method on your drawing by techniques in Fig. 6-7, C, D, and E.

Wrapping is done by a special wire-wrapping tool. Large wire-wrapping jobs will be done by automatic wrapping machines. The wrapping tools strip the wire's insulation and then wrap the wire tightly around the wrapping post.

Wire-wrapping has an economical advantage over soldering and crimp terminations. Wrapping can be set up much easier for automated machines.

Wire used for wrapping terminations is solid wire. The gages for this wiring method range from #20 AWG to #32 AWG.

POINT-TO-POINT WIRING DIAGRAMS

The purpose of a point-to-point diagram is to show the engineering, manufacturing, and service personnel the wiring between and across components. See Fig. 6-8. Point-to-point diagrams contain the information necessary to make or follow all wire connections. This wiring diagram can be shown on the

AMERICAN WIRE GAGE				
Size AWG	Diameter Inches	Area C/R MILS	Area Sq.In.	Weight LBS/ 1000 Ft.
10	.1019	10,380	.008155	31.43
12	.0808	6,530	.00513	19.77
14	.0641	4,110	.00323	12.43
16	.0508	2,580	.00203	7.818
18	.0403	1,620	.00128	4.917
20	.0320	1,020	.000804	3.092
22	.0250	640	.000503	1.945
24	.0201	404	.00317	1.223
26	.0159	253	.000199	.7692
28	.0126	159	.000125	.4837

Fig. 6-5. The American Wire Gage table.

OHMS AND CURRENT RATINGS		
Wire Size AWG	Maximum Current Amperes	Ohms Per Lb.
10	50	.031
12	40	.080
14	30	.203
16	20	.513
18	15	1.296
20	11	3.280
22	9	8.290
24	5	20.900
26	2	52.900
28	.5	133.900

Fig. 6-6. Current ratings for wires. Military standards allow only 60% of these current values.

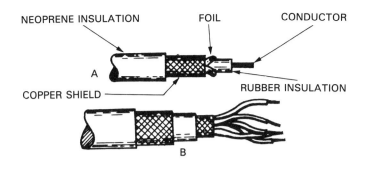

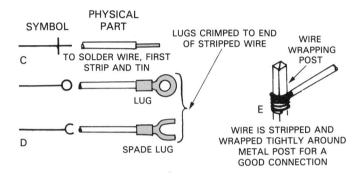

Fig. 6-7. A—Coaxial cable. B—Multi-conductor shielded cable. C, D, E—The three ways wires are terminated. Letter E shows wire wrapping post and wrapped wire.

assembly drawing. The diagram on the assembly will be included only if it is practical, and if room is available. The point-to-point drawing will not have a parts list. All the necessary items will be called out on the assembly document.

Some point-to-point diagrams show wiring paths on a background of components which are not drawn to scale. See Fig. 6-8 again. The components are drawn out of scale to fit the requirements of a very complex wiring diagram. The wiring is easier to follow when wires are on equal spacing. A second drawing often shows the components drawn to the true scale.

Point-to-point diagrams show the general physical arrangement of the component parts, Fig. 6-9. General rules for wiring diagrams are:
1. Minimize jogs in lines. Refer to Fig. 6-9 again for an example.
2. Run lines with a minimum of crosses.
3. Space lines a minimum of 3/8 in. apart.
4. Separate every three or four lines with an extra wide spacing when groups of lines run parallel to each other. This helps the reader's eye follow the individual lines.

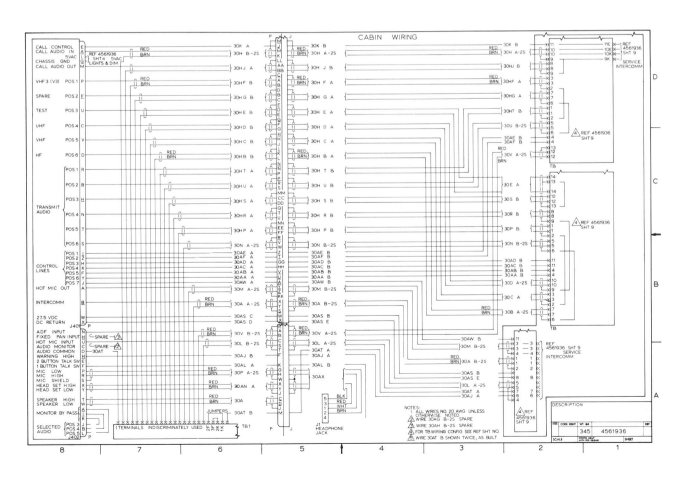

Fig. 6-8. A complex industrial point-to-point drawing shows assembly specifications.

5. Label components on the right side. This will help the reader when searching over a large drawing to find a specific component.
6. Letter the components with larger bold letters. Use smaller lettering for internal terminals.
7. Number components from the upper left hand corner. Make the diagram read as a book with the highest component number being in the bottom right corner.

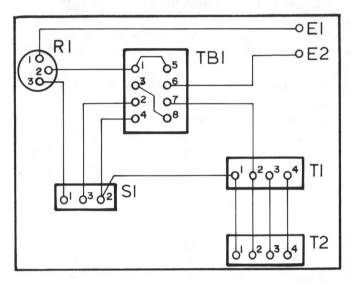

Fig. 6-9. A correctly drawn point-to-point wiring diagram. Note the misnumbering of TB1 and S1 to keep from crossing and jogging. This is a good practice.

PICTORIAL POINT-TO-POINT DRAWING

Occasionally when there is a simple point-to-point drawing to be made, it can often be drawn as a pictorial. Fig. 6-10 is a good example of a pictorial point-to-point drawing. However, pictorials should only be attempted when there are only a small number of wires and simple chassis layouts.

HIGHWAY WIRING DIAGRAMS

The highway wiring diagram groups the wires together into major paths called highways, Fig. 6-11. The technique allows you to put many wires on a drawing because this organized method saves room. The drawing shows the physical arrangement of the component parts as we did in the point-to-point. It will be possible to tell each wire's destinations, color, and gage by looking at either of its ends.

In Fig. 6-11D we see a second method for showing highways. Note that some companies apply a

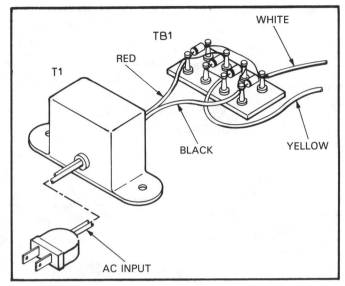

Fig. 6-10. A typical pictorial point-to-point.

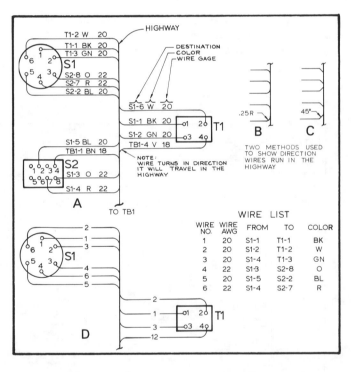

WIRE LIST				
WIRE NO.	WIRE AWG	FROM	TO	COLOR
1	20	S1-1	T1-1	BK
2	20	S1-2	T1-2	W
3	20	S1-4	T1-3	GN
4	22	S1-3	S2-8	O
5	20	S1-5	S2-2	BL
6	22	S1-4	S2-7	R

Fig. 6-11. A—Typical highway diagram. This method can handle many wires in an organized manner. B, C—Two methods for routing individual wires into the highway. They both show direction of travel. D—An alternate method for highway diagrams. It uses a table to show the wiring information. The table may be typed on the drawing, reducing drafting time. But it is slower to read than in Part A because the reader goes between the table and drawing.

number to each wire and then create a separate table. The wire number in the table will supply destination, color, and gage.

BASELINE DIAGRAMS

Baseline diagrams are like highway diagrams in two ways. They both can handle many wires in an organized manner, and they bundle the wires together in one main line, Fig. 6-12. They also have a couple of differences. One is the placement of components. The highway diagram is very concerned with physical placement of the components. The baseline diagram just lines them up in a straight line.

Another difference is in the way the wires enter the main bundle. The highway diagrams shows which way the wire will be running in a bundle. The baseline just goes into the wire bundle at 90 degrees.

The method for drawing baseline diagrams is to:
1. Construct a light line across the middle of the paper.
2. Line up the component on either side of the line.
3. Take short lines from each component and run them into the center line at 90°.
4. Identify the wire destination and color.
5. Make the center line a dark bold line, Fig. 6-12.

Baseline drawings are used mostly for service manuals and maintenance books. They are especially good for this kind of information because they can neatly show many lines on a book size sheet of paper. These drawings will not normally be used for assembly work because the information is too limited.

INTERCONNECTION DIAGRAMS

Interconnection diagrams show the wiring between different electronics units and between subassemblies, Fig. 6-13. This document is similar to the point-to-point wiring diagram. Each cable assembly and electronic unit will be called out and assigned a title and drawing number. Note the subassemblies are shown in phantom lines. Internal connections of electronic units are not shown.

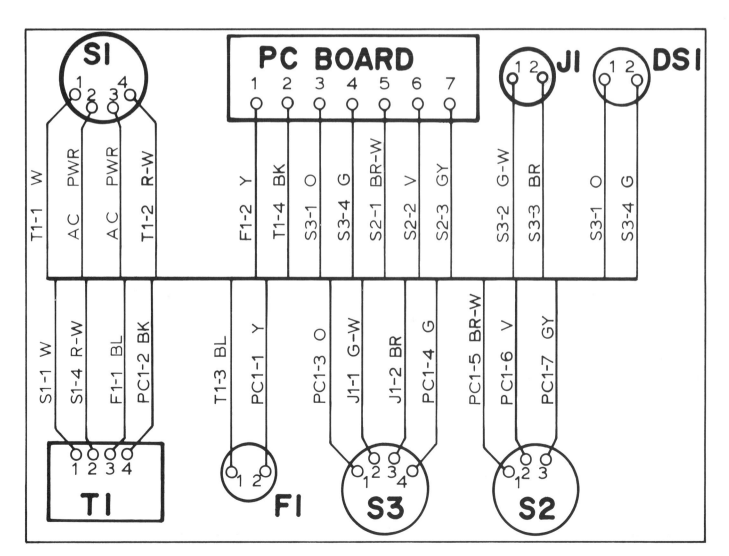

Fig. 6-12. A baseline diagram showing another method of controlling many wires in an organized manner.

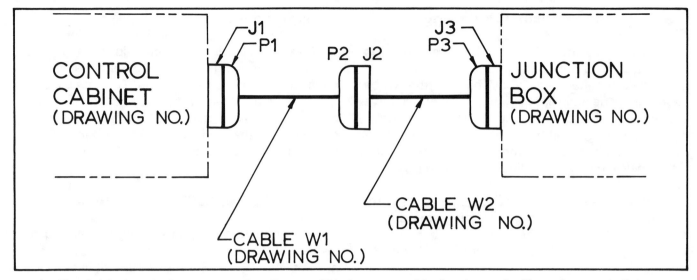

Fig. 6-13. A typical interconnection diagram. This is an assembly drawing and will require a parts list. Subassemblies on an interconnection diagram are shown in phantom lines.

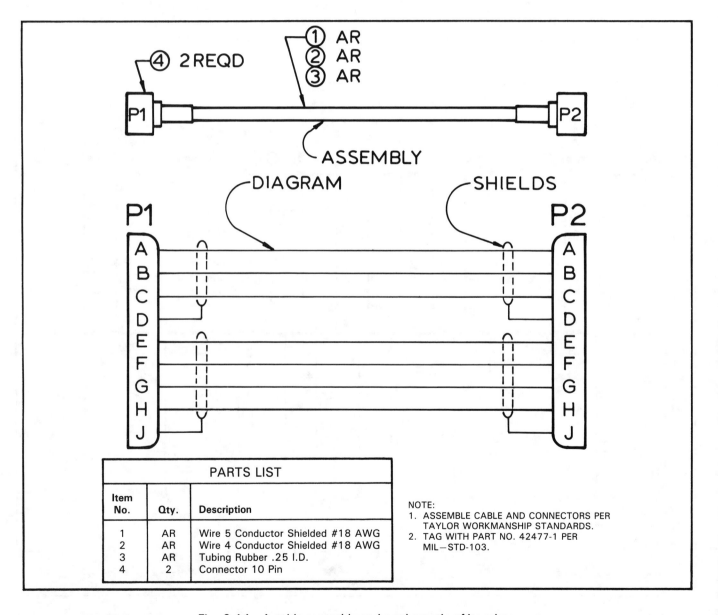

PARTS LIST		
Item No.	**Qty.**	**Description**
1	AR	Wire 5 Conductor Shielded #18 AWG
2	AR	Wire 4 Conductor Shielded #18 AWG
3	AR	Tubing Rubber .25 I.D.
4	2	Connector 10 Pin

NOTE:
1. ASSEMBLE CABLE AND CONNECTORS PER TAYLOR WORKMANSHIP STANDARDS.
2. TAG WITH PART NO. 42477-1 PER MIL—STD-103.

Fig. 6-14. A cable assembly and a schematic of its wires.

CABLE ASSEMBLY DRAWINGS

Cable drawings are assembly drawings, Fig. 6-14. They contain all necessary information to manufacture a finished cable. The drawing will include the following information:
1. A complete parts list.
2. A drawing showing all components.
3. Reference designations for each component.
4. A wiring diagram most often is part of the drawing. It will show the internal wires in the cable.
5. A general note section which will guide the assembler through the assembly.

WIRE (CABLE) HARNESS ASSEMBLY DRAWING

The wiring harness is the only wiring drawing drawn to exact scale, Fig. 6-15. It is drawn to scale because it is not just a drawing, but it is also a tool.

This tool will be used in manufacturing so that many identical parts can be created. Harness drawings will be supported by a wiring list and parts list. It is an assembly so it will need to be supplemented with all information needed by the assembler. The benefits of this drawing are:
1. It will support high volume manufacturing.
2. It will not require high priced technicians.
3. Quality control of the wiring is easier.
4. The assembly is less expensive to produce than many separate wires.

Before we can begin a harness drawing, we must know the exact placement of all the electrical components to be hooked up. The layout of the drawing and the routing of the harness will be decided by studying this arrangement, Fig. 6-16. Once we know where the harness will run, we can plan the layout on the drawing.

Routing of the wires in a harness is accomplished

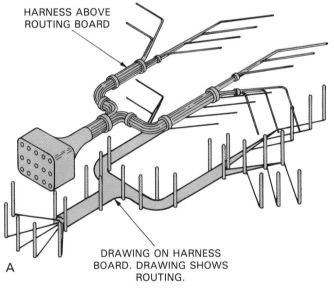

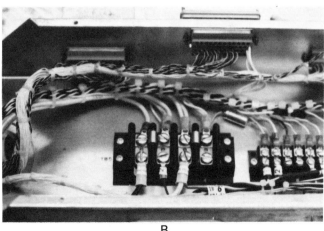

Fig. 6-15. A—Harness being removed from the harness routing board. B—Picture of harness installed in equipment.

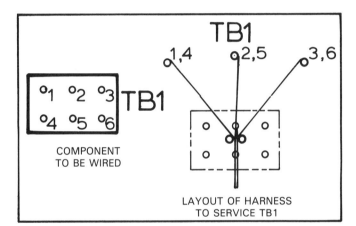

Fig. 6-16. Terminal block one needs to be wired. The right view shows how the wiring will be routed to service TB1. Note: The wires go beyond the position of TB1. This is to provide a service loop so the wires can be easily hooked and unhooked.

by retaining them between HARNESS POSTS, Fig. 6-17. Harness posts will be driven into a routing board as the drawing directs. Harness posts will also be used as securing posts for each end of the wire. The wire will be wrapped around the starting post. Once secured, it will be run through the routing posts as described by the wire list. After routing, it will be secured around an ending post.

LACING or strapping of the harness will be done after all the wires are routed. See Fig. 6-18. Lacing or strapping is the bundling of the wires into a permanent unit. Once the wires have been permanently bundled, they can then be lifted up off the routing board. After a harness has been removed,

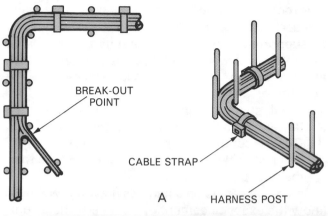

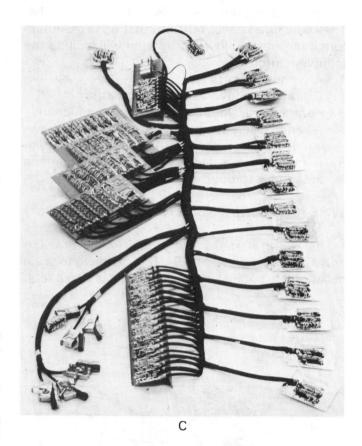

Fig. 6-17. A—Wiring being routed through harness posts. B—Picture of a wiring harness being automatically routed.
C—Example of a completed wiring harness. (Amphenol North America Division, Bunker Ramo Corp.)

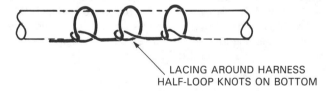

Fig. 6-18. An example of lacing being applied around a wire bundle. See Fig. 6-17 for an example of cable straps. Lacing and cable straps are applied to keep the wire bundle in the desired shape.

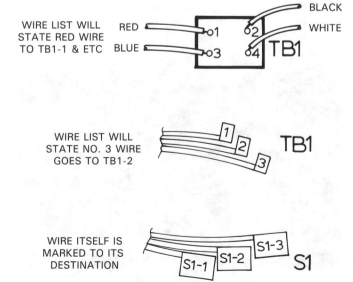

Fig. 6-19. Three ways to identify wires to aid in the harness installation.

another duplicate harness may be started.

After the harness has been manufactured, it will go to a higher assembly level where it will be installed. In order to make the installation easier, we identify each wire in the bundle. There are three methods of identifying wires, Fig. 6-19. The three methods are: Color, number, or destination. When using colors, each wire carrying a different electronic signal will have a different color. Numbered wires will be numbered on both ends with the same number tag. A wire identified by destination, will have the exact place where it is to be terminated tagged right on its ends. The destination method will eliminate the need for a wire list during installation. Numbered and colored wires must have a wire list with the harness in order to complete installation.

TYPICAL NOTES USED ON WIRING DIAGRAMS

Here are often-used notes for wiring diagrams:
1. This drawing used with—
   Assembly drawing _____
   Schematic drawing _____
   Wiring diagram   _____
2. Wire lengths determined by prototype
3. Wire color coding per MIL-STD-681
4. Wiring must conform to _____
5. Soldering will conform to _____
6. Unless otherwise specified, all wires are ____ _____
7. Lace harness at each breakout point and every _____ inch in between
8. Apply cable straps at each breakout point and every _____ inch in between

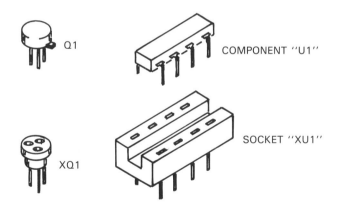

Fig. 6-20. Components and sockets and their reference designations.

REFERENCE DESIGNATIONS

Reference designations shall be identical to those on the schematic except for component sockets which are prefixed "X". See Fig. 6-20.

COMPONENT REPRESENTATION

Component representation shall be just a physical outline suggestive of the component's features, Fig.

6-21. This is meant to be a simplified view as seen from the wiring side.

TERMINAL IDENTIFICATION

Each terminal must be identified. Most components and connectors are marked adequately, but if not: sufficient details must be supplied with a wiring diagram. Leads of components, such as transistors, diodes, electrolytic capacitors, batteries, and other devices shall have their terminals identified or polarity marked, as in Fig. 6-22.

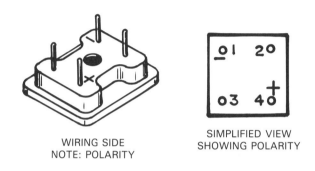

Fig. 6-21. The actual component, left, and its representation on the right. Note: Pins have been assigned numbers. This helps the technician during wire installation.

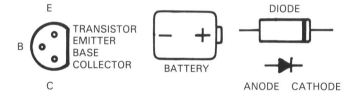

Fig. 6-22. Polarized components with their leads identified.

REVIEW QUESTIONS

1. What three things are considered when we choose a wiring method?
2. What information is normally included in the wiring list?
3. Name two types of wires.
4. What is the advantage of stranded wiring?
5. What does 7/26 mean when we relate it to wiring?
   a. 7 strands wrapped around 26.
   b. 7 gage wire wrapped by 26 gage wire.
   c. 7 strands of 26 gage wire.
   d. 7 strands that total 26 circular mills.
6. Gage of wire is determined by the _____ _____ _____ Standard.
7. What is affected by wire length and diameter?

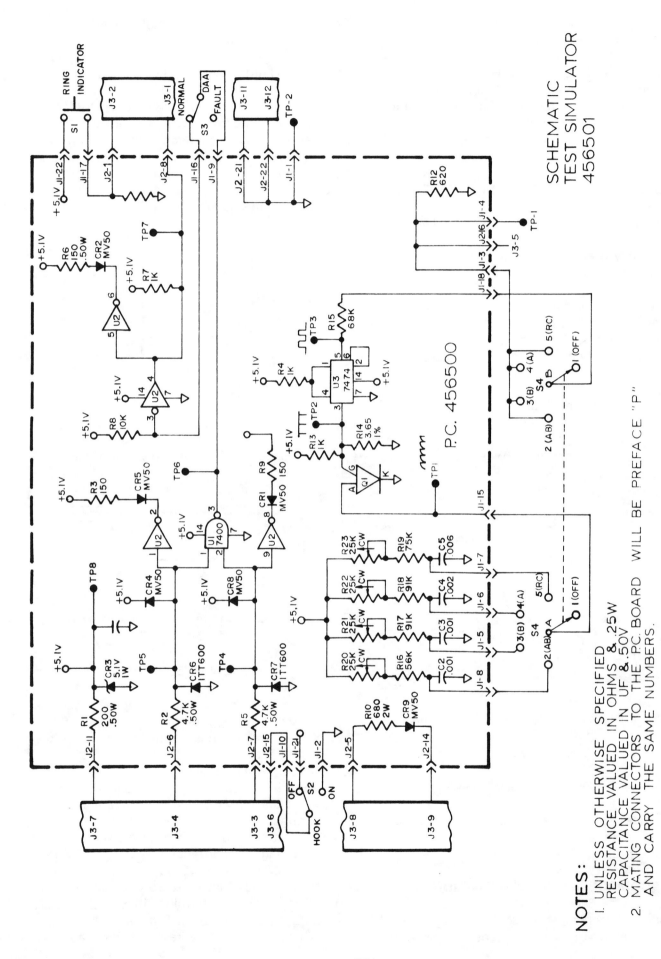

Fig. 6-23. A printed circuit board, enclosed in dashed lines, and the connecting lines to switch, test points, and an external connector, J3. Create a wiring list and parts list.

SCHEMATIC TEST SIMULATOR 456501

NOTES:

1. UNLESS OTHERWISE SPECIFIED RESISTANCE VALUED IN OHMS & .25W CAPACITANCE VALUED IN UF & .50V

2. MATING CONNECTORS TO THE P.C. BOARD WILL BE PREFACE "P" AND CARRY THE SAME NUMBERS.

8. How do we use bus wire?
9. Shielded or _____ cable reduces interfering radiation.
10. A point-to-point diagram shows the physical arrangement of the components. What other information does it convey?
11. When would we use pictorial point-to-point diagrams?
12. What is the advantage of a highway diagram?
13. Cable assembly drawings will contain what information?
14. Why is the cable harness assembly drawn to scale?
15. What are the advantages of cable harnesses?
16. How are harness posts used?
17. Cable straps or lacing serve what function?
18. What determines the type of termination a wire will make?
19. The most economical termination method with automation is _____ (soldering, crimping, wrapping).

20. Subassemblies on _____ diagrams are shown in phantom lines.

PROBLEMS

PROB. 6-1. Using the test simulator schematic, Fig. 6-23, create a wiring list and parts list. Wiring between the printed circuit connectors, switches, test points, and connector needs to be listed. See Fig. 6-2 for wire list format. Wiring shorting out the switches has been accomplished in a subassembly so do not list.

PROB. 6-2. Draft a point-to-point wiring diagram for Fig. 6-24, the test simulator. Check Fig. 6-25 for component numbering and sizes. Fig. 6-26 will show panel mounting positions. Use the wiring list generated in Problem 1 to aid in this problem. If the wiring list was not completed use the information indicated in

Fig. 6-24. An exploded view of a test simulator package. This is the final assembly drawing. Draft a point-to-point diagram.

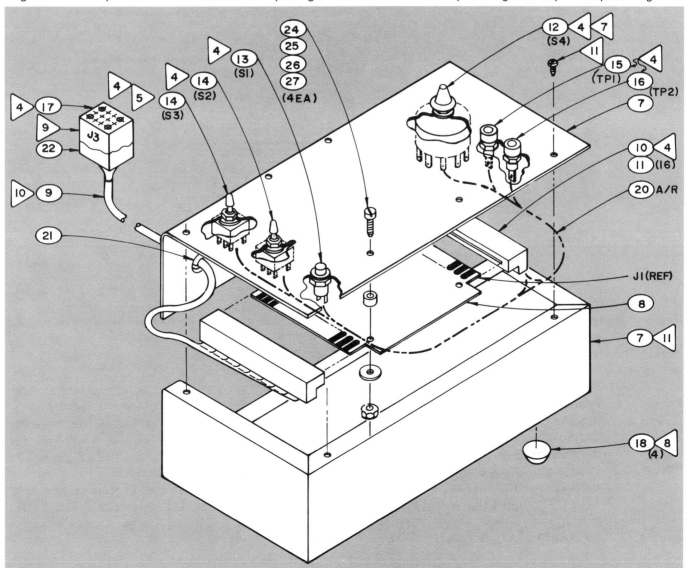

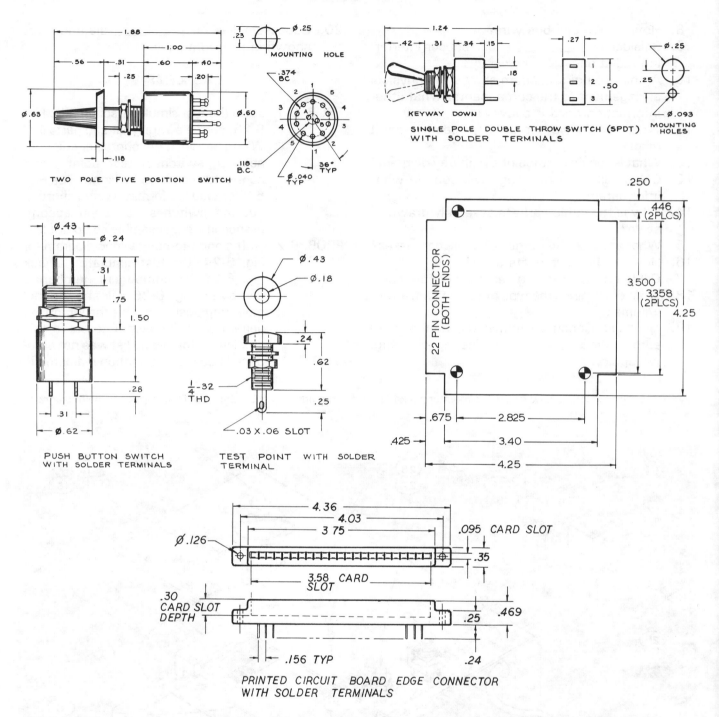

Fig. 6-25. These components used in the test simulator project are important for planning the case mounting hardware. The chapter problems refer several times to this Figure.

Problem 1. Start by "undoing" the assembly in Fig. 6-11A. All wire lengths determined at assembly. Note: Wiring is installed with panel turned upside down. See Fig. 6-27.

PROB. 6-3. Create a highway diagram using information described in Problem 1 and 2. Fig. 6-27 shows the layout paths for the highway. Use as much information as you can from above problems.

PROB. 6-4. Using all the accumulated information from above problems, create a wire harness drawing. Note: This drawing is drafted to a 1/1 scale. It will be a tool for manufacturing. Check Figs. 6-23, 6-24, 6-25, and 6-26.

PROB. 6-5. Draft a baseline diagram of the test simulator. Use example in Fig. 6-12, for example. Assign a color to each wire.

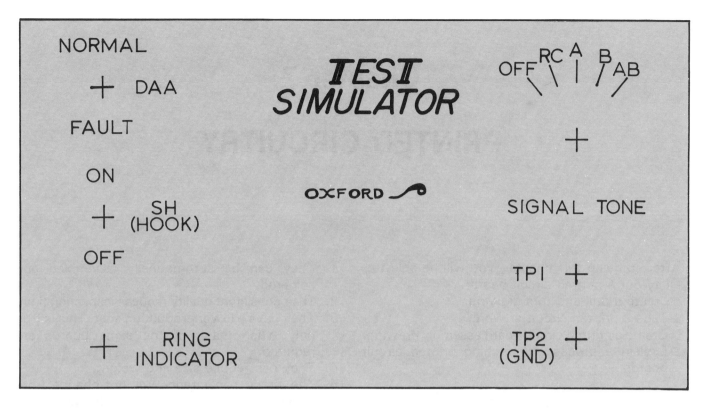

Fig. 6-26. The front panel of a test simulator. The switch and test point positions are shown full scale. This is a silk-screen artwork for the panel. Some of the information is needed in several chapter problems.

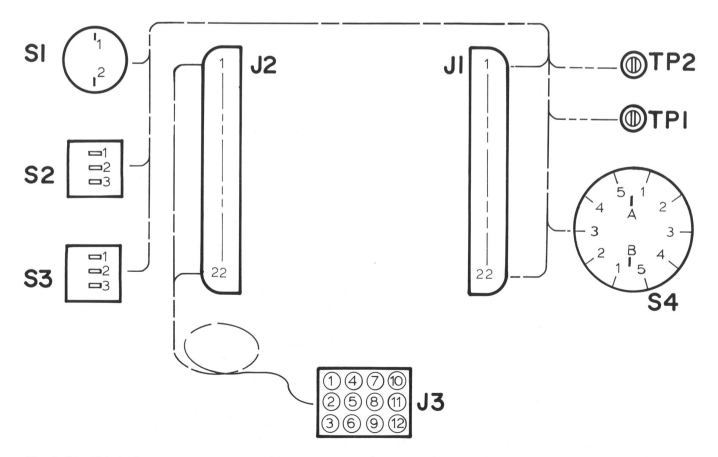

Fig. 6-27. This is the basic layout of the highway diagram for the test simulator. Complete this highway diagram using wire destinations given on the schematic.

Chapter 7

PRINTED CIRCUITRY

After studying this chapter, you will be able to:
- Lay out a printed circuit board.
- Create a drill and trim drawing.
- Tape a printed circuit artwork.
- Create a printed circuit board assembly drawing.
- Explain a process for etching printed circuit boards.

The drafter will begin creating the printed circuit board (PCB) after receiving a schematic or logic drawing and a list of components. The engineer will provide this information and any other necessary design data.

SCHEMATIC OR LOGIC DIAGRAMS

These drawings will show the functions and major circuit paths of the electronics circuit. See Fig. 7-1. The schematic shows the power and signal paths for individual components. The logic drawing shows the data paths for digital or analog circuitry. The first step in PCB layout and design is to study the engineering input.

DESIGN DATA

The engineer will provide the following design information:
1. The voltage and amperage ratings of the circuit.
2. Heat-sink requirements for heat-dissipating components.
3. Grounding requirements.
4. Board size and mounting technique.

Numbers 1, 3, and 4 of these design requirements will be discussed later in this chapter.

ADVANTAGES OF PRINTED CIRCUITS

Printed circuitry has some distinct advantages over other methods of interconnecting circuits. The major advantages are:

1. They can be automatically assembled and soldered.
2. Their consistent quality reduces inspection time.
3. They have a lower production cost. This is true only when many of the same boards are required.
4. They can reduce size and weight.
5. The circuit capacitance will not change from circuit to circuit as in wiring techniques.
6. Their plug-in capability makes them easier to service.

The above advantages make printed circuits the heart of most electronic devices.

PCB LAYOUT

To start the layout we need to gather all the necessary component, circuit, and board information. The engineer will provide ratings (volts, amps), but you must gather much of your own component information. The necessary component information will be: the size of the component, how it is mounted on the board, and what size mounting holes are required, Fig. 7-2. With the gathered information, you can now start to work on the PCB.

DESIGN SCALE

The most common scales used to create Printed Circuit Artworks are 1X, 2X, and 4X. All complex circuitry should be prepared at an enlarged scale. Drafting errors such as poor placement of mounting pads, conductor spacing, and other errors will be reduced by the multiplier when photographed to give a 1 to 1 scale. Precision layout grids, manufactured component patterns, numerically controlled hole drilling, and automatic component insertion dictate the artwork scale. NC drilling and automatic stuffing should be at .025 in. intervals. Therefore, the precision grids are .10 or .05 and the component patterns are designed around that interval.

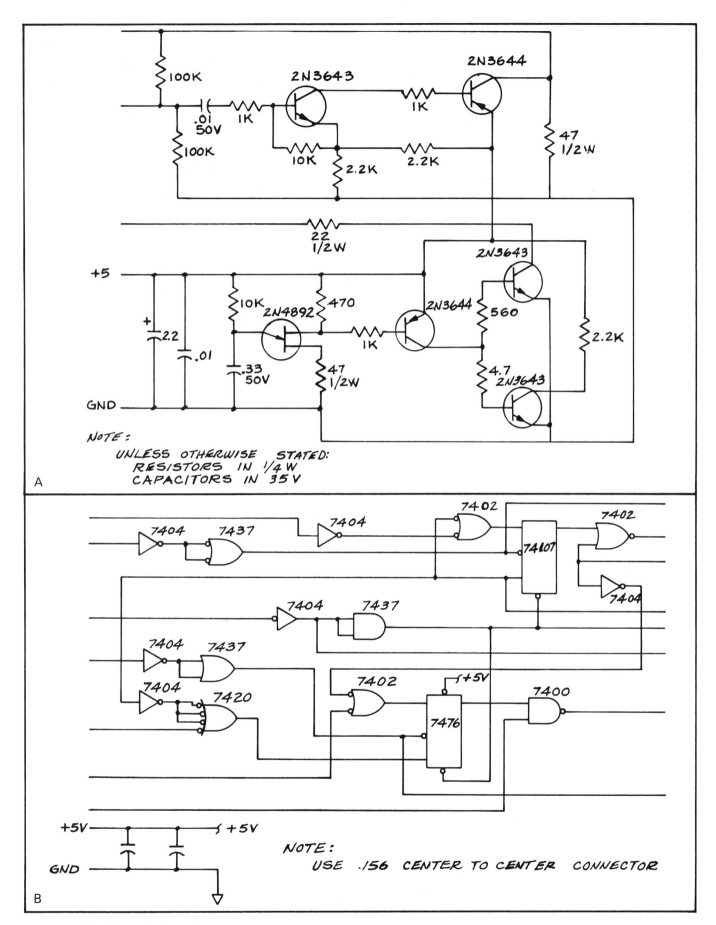

Fig. 7-1. A—The schematic interconnection information. Component information is included for size consideration. B—The same information on a logic drawing.

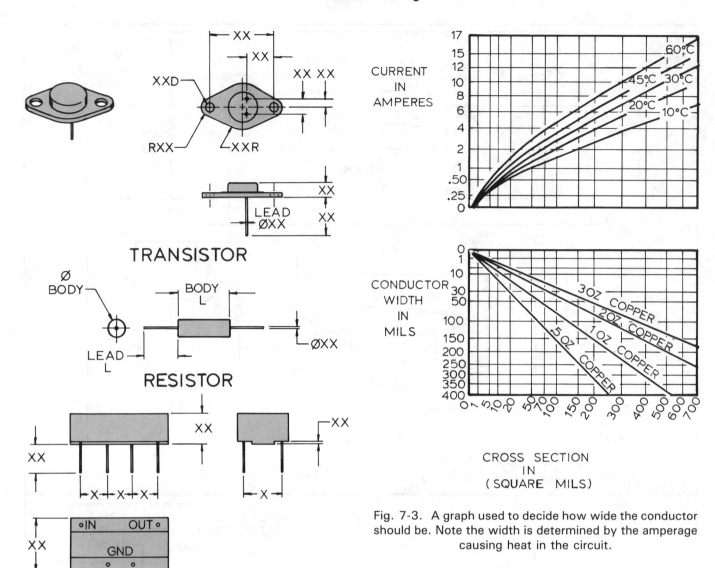

TRANSISTOR

RESISTOR

A DIRECTIONAL COMPONENT

Fig. 7-2. This is an example of the information you will find in a components catalog. This data is vital to the completed design layout.

CROSS SECTION
IN
(SQUARE MILS)

Fig. 7-3. A graph used to decide how wide the conductor should be. Note the width is determined by the amperage causing heat in the circuit.

Artwork with a 1 to 1 scale is used only when you need a quick prototype board, or limited accuracy is required.

CONDUCTORS

Care must be taken in deciding width and spacing of conductors. Conductor widths or cross sectional area is determined by current carrying requirements. Fig. 7-3 shows the ratio of amperes to the cross sectional area of the conductor.

Spacing of conductors is dictated by the amount of voltage in the circuit, the altitude when used, and the board's coating. When the board is used at higher altitudes as in an aircraft, you are required to increase the voltage gap. The lower barometric pressure at higher elevations will allow the current to arc across the voltage gap easier. Coating the board with an insulation helps prevent voltage arcing. See Fig. 7-4 for spacing requirements.

BOARD SIZE AND STYLE

The simplest board to manufacture is a single layer board, Fig. 7-5. This board will have circuitry only on one side. The opposite side will be the component side. Single layer boards are the least expensive to manufacture. However, they will not meet all requirements. High density boards require more layers. They require the use of double-sided or multilayer boards.

Double-sided boards allow you to apply circuitry to both sides of the board, see Fig. 7-6. With this method, you will try to apply the maximum amount of circuitry on the circuit side. Having less circuitry on the component side will reduce the problem of having components over the circuitry. In extremely

CONDUCTOR SPACING			
VOLTAGE BETWEEN CONDUCTORS (DC OR AC) PEAK VOLTS	SEA LEVEL TO 10,000FT	10,000FT AND ABOVE	COATED BOARDS ANY ALTITUDE
0 - 30 0 - 50 0 - 150	.025	.026	.010
31 - 50 51 - 100 151 - 300	.050	.062	.015
51 - 100 101 - 170 301 - 500	.100	.125	.020
101 - 300 171 - 250 500 & ABOVE	.0002/VOLT	.250	.030
301 - 500 251 - 500		.500	.060
500 & ABOVE 500 & ABOVE		.001/VOLT	.00012/VOLT

Fig. 7-4. This table shows how voltage, elevation, and an insulation coating can determine the spacing of conductors.

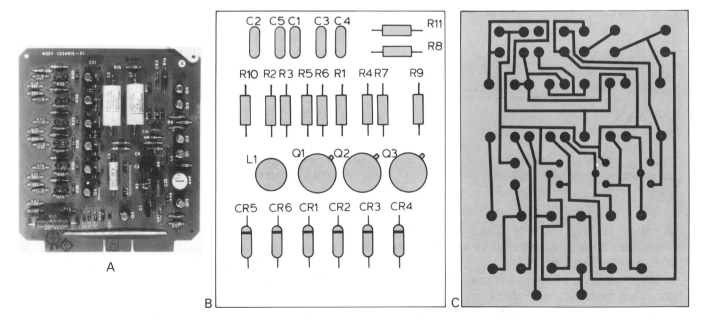

Fig. 7-5. A—Single-sided board. It is the least expensive and simplest to manufacture. B—Marking artwork, single-sided board. C—Circuit side of board in B.

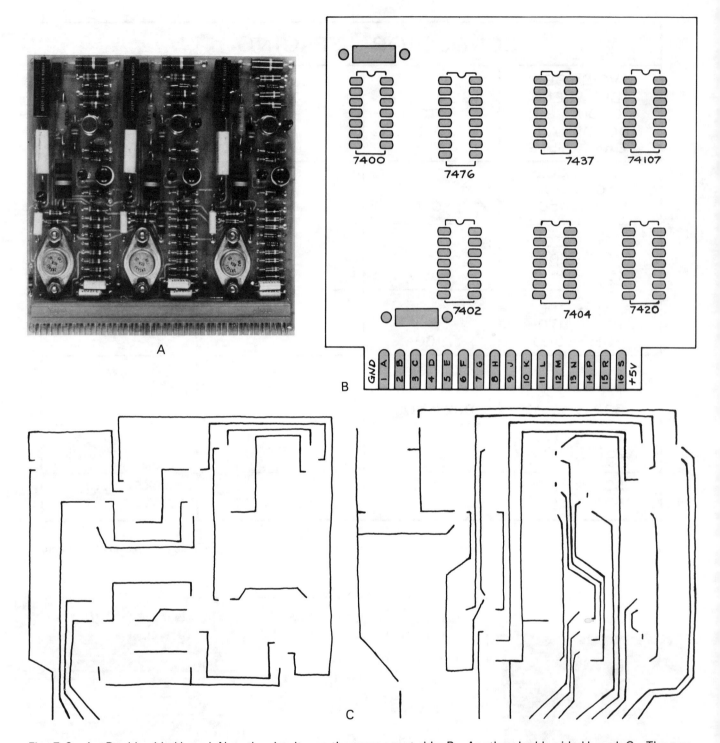

Fig. 7-6. A—Double-sided board. Note the circuitry on the component side. B—Another double-sided board. C—The conductors normally flow horizontal on one side of the board and vertical on the other. Sketches refer to board in B.

dense circuity, you may use multilayer boards. Multilayer boards with as many as 20 layers have been used. See Fig. 7-7.

Once the style of board has been chosen, you can consider board size. Experience will help when choosing the size of board. Until you gain experience, you may need to look at previously manufactured boards for a guide.

A SKETCH FOR A LAYOUT

In designing a circuit board, it is better to do a quick sketch. The sketch will allow you to get an idea of how the circuit will flow. This sketch is not intended to be a work of art. It only needs to be organized enough so that another drafter can work from it if required, Fig. 7-8.

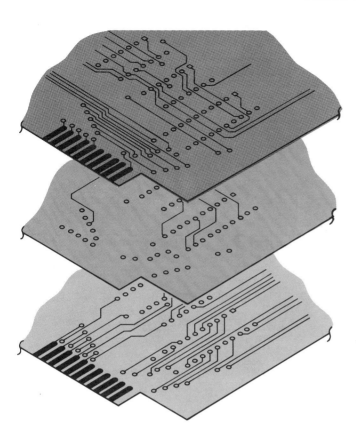

Fig. 7-7. The levels of a multilayer board. Boards have been manufactured with up to 20 layers.

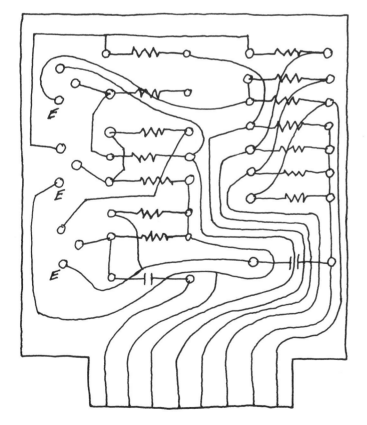

Fig. 7-8. A sketch of the PCB used to decide on component placement and how the interconnections will flow.

PREPARE FORMAL LAYOUT

After you achieve a workable sketch, you are ready to go to a more formal layout. This layout will be accomplished with templates or PUPPETS®. With templates, you can draw the scaled component outlines on the layout. See Fig. 7-9.

Puppets are pre-made component outlines which are put on an adhesive backed format. They can be adhered wherever you wish on the layout.

This formal layout should be drawn on a gridded format. The first step will be to lay out the edges of the board, including any connectors. Second, you should apply DATUM TARGETS, Fig. 7-9. These

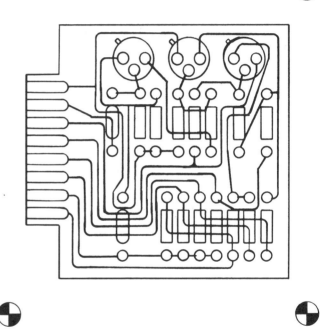

Fig. 7-9. A formal layout with each component shown true size and with all interconnections figured for taping.

targets will be used as exact points for measuring purposes. Three targets are normally applied. If more than one sheet will be used in the layout, each sheet should have targets. The targets will allow you to re-register the layers each time they are mated together. These targets will also be used for photographing the artwork and during board manufacturing. Photographic targets or reduction points may also be used to help in the etching process. Etching is the chemical removal of the copper on the circuit board. Fig. 7-10 shows how these points serve in the etching process when placed on the PCB.

Fig. 7-10. A target being used for checking the etching process. A—A good etch. B—Needs more material etched away. C—Over-etched.

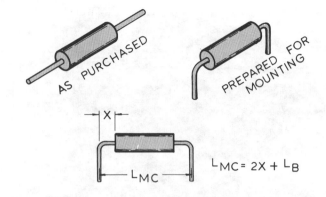

After you have finished the above tasks, you can continue with the following:

1. Represent the physical shape of each component on the layout, Fig. 7-11.
2. Indicate the polarity of polarized components, Fig. 7-12.
3. Establish the mounting requirements for each component, Fig. 7-13.
4. Space and arrange components correctly. See Fig. 7-14.
5. Locate adjustable components so they can be adjusted easily, Fig. 7-15.
6. Locate test points so they can be probed with

$L_{MC} = 2X + L_B$

L_{MC} = LENGTH OF CENTER TO CENTER DISTANCE
L_B = LENGTH OF COMPONENT BODY
X = LEAD LENGTH FROM BODY

"X" IS DETERMINED BY USING THE FOLLOWING TABLE:

LEAD DIA	X (MIN)
.015	.050
.020	.075
.030	.090
.040	.115
.050	.120

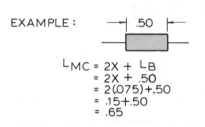

EXAMPLE:

$L_{MC} = 2X + L_B$
$= 2X + .50$
$= 2(.075) + .50$
$= .15 + .50$
$= .65$

Fig. 7-13. An example of how to figure mount spacing on axial lead components.

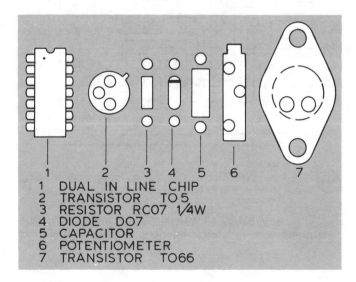

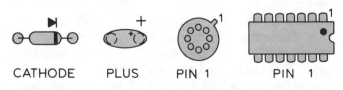

1 DUAL IN LINE CHIP
2 TRANSISTOR TO 5
3 RESISTOR RC07 1/4W
4 DIODE DO7
5 CAPACITOR
6 POTENTIOMETER
7 TRANSISTOR TO66

Fig. 7-11. An example of how components would be represented on a layout drawing for a PCB. The drafter will determine circuit path locations from terminals.

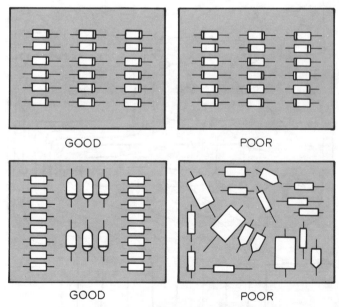

Fig. 7-14. Arrange components for ease of assembly and inspection. Note how much longer it would take to inspect correct installation of the poor arrangements.

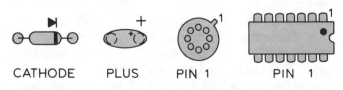

CATHODE   PLUS   PIN 1   PIN 1

Fig. 7-12. Indicate polarity on components that can be plugged in incorrectly.

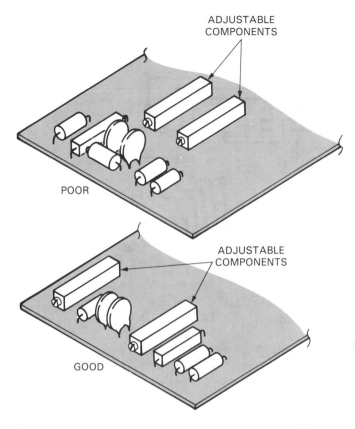

Fig. 7-15. Place adjustable components so the adjusting tool can easily reach them.

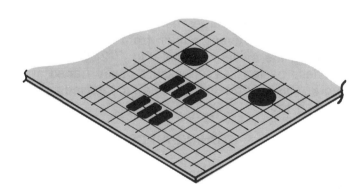

Fig. 7-16. Locate all test points so that measurements can easily be made with equipment in operation.

Fig. 7-17. The placement of pads on grid crossing. This is done to aid automated drilling and component assembly.

the unit operating, Fig. 7-16.
7. Place all mounting pads if possible on grid line crossings, Fig. 7-17.
8. Indicate size of holes required to mount components, Fig. 7-18.
9. Select terminal size required to solder component to PCB, Fig. 7-19.

RECOMMENDED DRILL HOLES FOR COMPONENT LEADS		
LEAD SIZE (DIA)	UNPLATED HOLE DIA	HOLE DIA FOR PLATING
.0063 TO .0113	.014 ±.003	.020 ±.005
.0126 TO .0226	.028 ±.003	.030 ±.005
.0253 TO .0320	.040 ±.004	.040 ±.006
.0359 TO .0453	.052 ±.004	.052 ±.006
.0508 TO .0571	.062 ±.004	.067 ±.007

Fig. 7-18. A table of recommended hole sizes for different component lead diameters.

SELECTING TERMINAL SIZE		
LEAD DIA	TERMINAL DIA	
.0063 TO .0113	.065	COMPONENT LEAD DIA
.0126 TO .0226	.075	
.0253 TO .0320	.090	
.0359 TO .0453	.100	MOUNTING TERMINAL
.0508 TO .0571	.115	

Fig. 7-19. A table showing the mounting pad diameter for the different lead diameters. This is to insure an adequate solder area around each terminal.

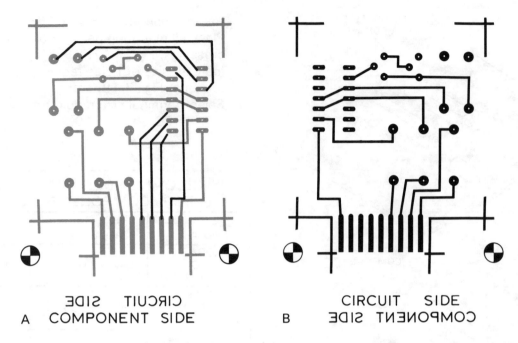

CIRCUIT SIDE
A   COMPONENT SIDE

B   CIRCUIT SIDE
COMPONENT SIDE

Fig. 7-20. A—The blue tape side on a single format artwork. B—The format is flipped over showing the red tape side of the circuit.

PRINTED CIRCUIT ARTWORK

Artwork can be created in different ways. There are two basic methods used in most companies. One method uses red and blue tape on a single sheet of plastic format. The second method is to use different sheets of format for each board level.

RED AND BLUE ARTWORK

Using the red and blue method, you can create a double-sided PCB on a single artwork sheet. For the double-sided board, start by placing component mounting pads on a clear film. After placing the pads, continue by taping the blue side (usually the component side). See Fig. 7-20A. Tape all interconnections on this side, then turn the film over and tape the solder side, which is usually done with red tape, Fig. 7-20B.

Do not tape both red and blue circuitry on one side of the film. This error will make it difficult to make changes on interconnections underneath other colored tape.

MULTI-SHEET ARTWORKS

Multi-sheet drawings will start with placement of component mounting pads on a single layer of film (the pad-master), Fig. 7-21. Then another film will be placed over the pad-master and the component side will be taped. Each additional layer will be placed over the pad-master and taped.

TAPING TECHNIQUES

The taping assignment is a critical job. It is one time when you are not just creating a drawing but also a manufacturing tool. Many times you are required to tape the artwork to ±.005 in. Consider five basic taping rules:

1. Avoid sharp corners, which can cause problems in etching, soldering, and with foil delamination, Fig. 7-22A.
2. Keep interconnections as short as possible. See Fig. 7-22B.
3. Keep the tapes and media clean. Any dirt or smudges will show up on the artwork photos.
4. Spread the circuitry out so it is evenly placed throughout the board. If it is concentrated in one area of the board, it will cause localized heating when automatic soldering. This uneven heating will warp the PCB.
5. Make terminals larger than interconnecting lines. This will prevent solder from flowing away from the solder terminal, Fig. 7-22C and D.

These basic rules will assure quality taped artwork. This artwork is then ready to be used as the etching tool.

ETCHING A PCB

Most industrial PC boards are created by using a photographic process. You will study this process to find how your artwork will be used in PC board manufacturing. The steps in manufacturing normally

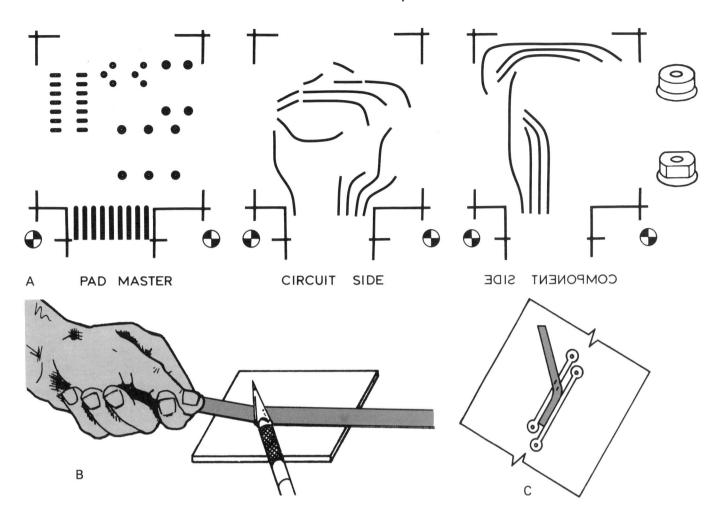

Fig. 7-21. A—The taping of a multi-sheet artwork. Note registration pins. B—Overlaptape on pad. Cut by lifting tape against fixed blade to avoid cutting the taped circuit paths below. C—Space the conductors with temperary spacing tape.

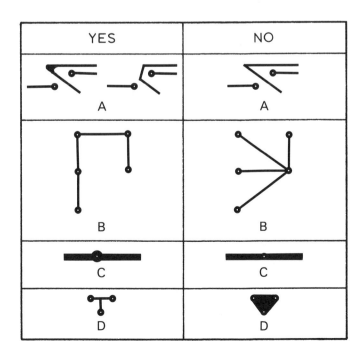

	YES		NO
	A		A
	B		B
	C		C
	D		D

Fig. 7-22. General taping rules for all PC boards. Some companies add their own specific requirements.

follow this standard sequence:
1. Purchase a PCB, Fig. 7-23, which is made up of three layers:
   a. The board, a flat stiff sheet of insulating material, often called the substrate.
   b. Copper foil laminated to the substrate.
   c. A photosensitive resist that will harden when exposed to ultraviolet light.
2. Drill holes from the drilling specifications. These

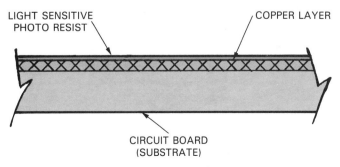

Fig. 7-23. The basic structure of a PC board.

specifications will be listed on the drill and trim drawing to be discussed later.

3. Expose the board to ultraviolet light. For this, use an artwork negative to control where the light exposes the board, Fig. 7-24.

4. Wash away unhardened resist and etch the board. The resist not exposed by the light will quickly wash away, allowing the etching chemicals to eat away the copper underneath. When the etching is finished, you will see the circuit you designed, Fig. 7-25.

5. Plating the connectors is necessary if they are pluggable edge connectors. The connectors are normally plated with gold because of its corrosion resistance and excellent current carrying ability. See Fig. 7-26.

6. Tinning is next. It is the coating of the PCB with a tin-lead alloy, Fig. 7-27. This is done to set the board up for soldering and to decrease the oxidation problem. Copper will oxidize more readily than tin-lead.

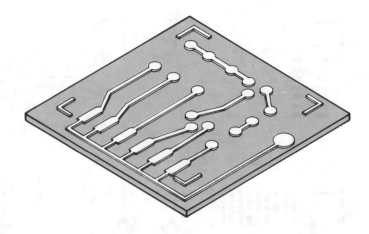

Fig. 7-25. The PCB after it has been etched by acids.

ULTRAVIOLET LIGHT SOURCE

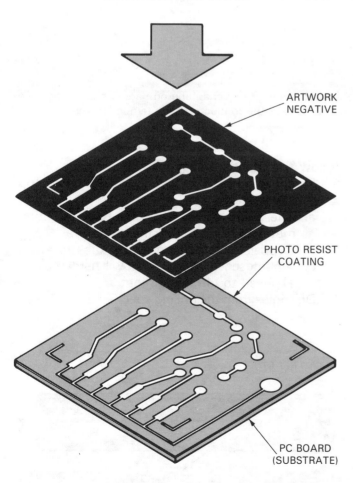

ARTWORK NEGATIVE

PHOTO RESIST COATING

PC BOARD (SUBSTRATE)

Fig. 7-24. The PCB is exposed to ultraviolet light. The light will harden the photo resist that it illuminates. This will set up the etching process.

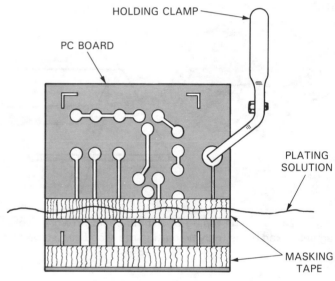

HOLDING CLAMP

PC BOARD

PLATING SOLUTION

MASKING TAPE

Fig. 7-26. The PCB has been masked off so the connecting contacts can be plated with gold.

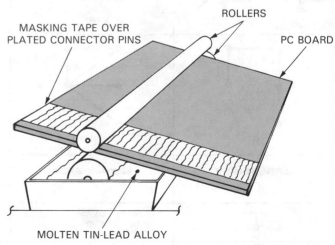

MASKING TAPE OVER PLATED CONNECTOR PINS

ROLLERS

PC BOARD

MOLTEN TIN-LEAD ALLOY

Fig. 7-27. Tin-lead alloy is being rolled on to the PCB. Note the previously gold plated connector contacts are covered with masking so they will not be coated with tin-lead.

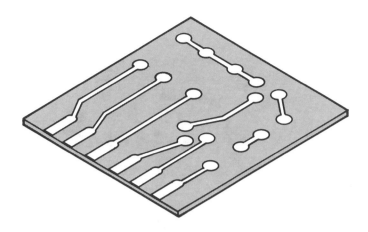

Fig. 7-28. The final board trimmed to size.

7. Trim the board to its specified final size. Fig. 7-28 shows the result.
8. Inspect the board for flaws. Flaws will be described in each company's PCB manufacturing specifications book.

PRINTED CIRCUIT DRAWINGS

There are at least five drawings required to document a PCB. They are: a logic or schematic drawing, a layout, artwork, drill and trim, and assembly with parts list. Optional drawings will be silkscreen drawings of component positions and component reference designations. You will study each of these drawings.

LOGIC OR SCHEMATIC

The schematic shows the circuit interconnections in the most compact way. It will be used to test and evaluate the completed circuit board.

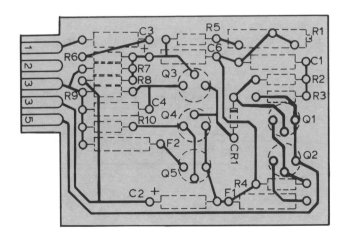

Fig. 7-29. A layout is an ink drawing meeting the physical requirements of the circuit.

LAYOUT

The layout is a preliminary ink drawing which is used to confirm design ideas for the electronic circuit, Fig. 7-29. It takes the symbolic schematic or logic drawing and transforms it into a model of the mechanical package. The layout will normally contain all the design information required to produce

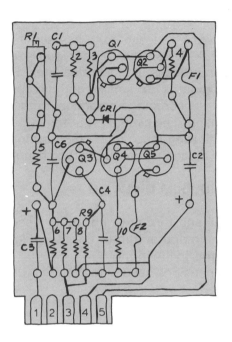

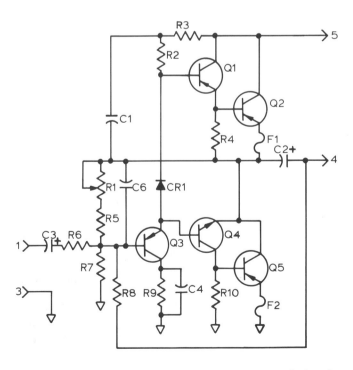

Fig. 7-30. A sketch as it relates to the schematic. A sketch may be the only step before the artwork. All interconnections are figured at this level.

a PCB. All interconnecting lines, components, component placement, component mounting, board size, and connectors should be considered at this drawing level.

Some PC board designs are created from a sketch, Fig. 7-30. The sketch may be the only step before the artwork. In other cases, it is just one of the steps before the artwork.

ARTWORK

The artwork is designed to be used as a tool to create the PCB. Because this drawing is used in the fabrication process, it is a critical step, Fig. 7-31. As the PC boards get smaller and the density gets greater, it is becoming impractical to tape the artwork by hand. The circuitry is automatically drawn by laser plotting machines. They are many times faster and much more accurate than hand methods.

DRILL AND TRIM DRAWING

The drill and trim drawing totally describes the design and manufacture of the PCB, Fig. 7-32. It includes the following:

1. All dimensions needed to cut the outline of the board.

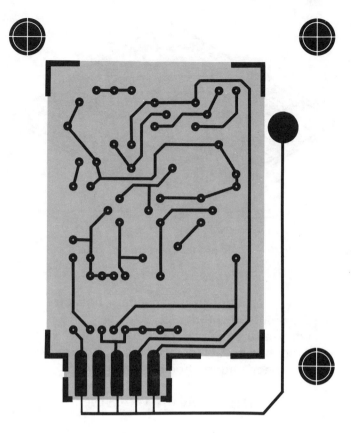

Fig. 7-31. The artwork for a single-sided circuit board. It is hand-taped.

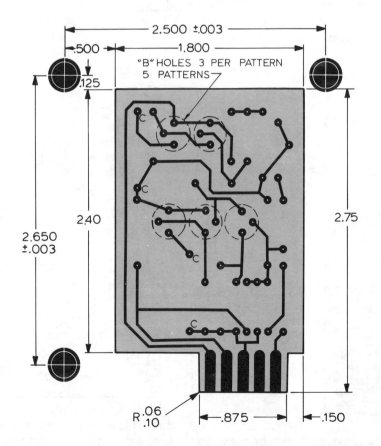

NOTE:
1. FABRICATE PER TANA WORKMANSHIP STANDARDS
2. ALL HOLES PLATED THRU
3. ALL HOLE DIMENSIONS APPLY AFTER PLATING
4. SILKSCREEN WITH ARTWORK 400500-1
5. GOLD PLATE CONTACTS PER MIL-G-45204 TO 70 MIL
6. BOARD MATERIAL TO BE .093 THK GLASS FILLED PHENOLIC
7. COPPER PLATING TO BE 2 OZ PER MIL-C-14550
8. DRILL ALL HOLES .040 EXCEPT THOSE LISTED IN TABLE I

TABLE I	
HOLE SYM	HOLE SIZE
B	.031
C	.052

Fig. 7-32. A drill and trim drawing. It is used to describe all the mechanical requirements of the PCB.

2. All drill hole information:
   a. Size of hole.
   b. Plated or unplated holes.
3. Description of material:
   a. Board material.
   b. Copper layer.
   c. Gold plating.
4. Photo reduction information.

ASSEMBLY DRAWING

The assembly drawing takes the etched board and adds all components and other assembly items, Fig. 7-33. It will show a view of the component side with all components stuffed. A complete parts list will identify all material and components needed. Notes will be added to clarify any information.

SILKSCREEN DRAWING OPTION

The silkscreen will normally show component reference designations and/or component place-

ment, Fig. 7-34. This information helps in board assembly, inspection, and service. It is vital when a board is set up to make optional choices. The options can be silkscreened on the board to make changes accurate and timely, Fig. 7-35.

REVIEW QUESTIONS

1. The logic drawing shows _____ paths for digital or analog circuits.
2. List some advantages of printed circuit boards.
3. Drafting errors are _____ after a drawing is reduced photographically.
   a. More obvious.
   b. Repairable.
   c. a and b.
   d. Less obvious.
4. Describe the etching process.
5. What factor does voltage play when you design a PCB?
6. List the information needed before you start a PCB layout.

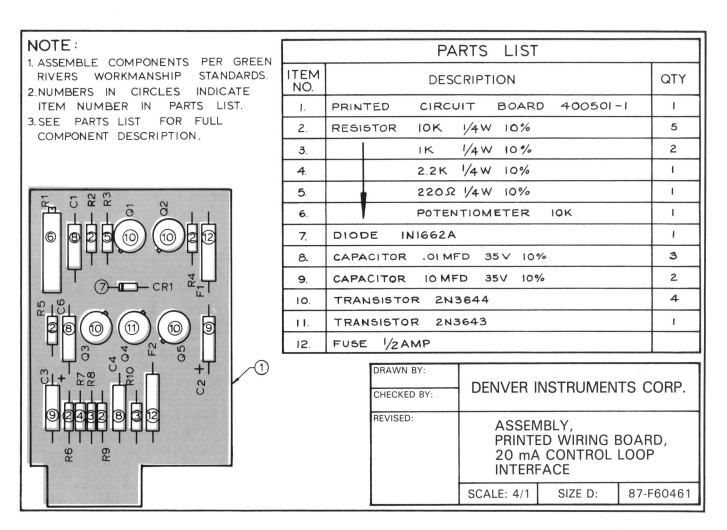

NOTE:
1. ASSEMBLE COMPONENTS PER GREEN RIVERS WORKMANSHIP STANDARDS.
2. NUMBERS IN CIRCLES INDICATE ITEM NUMBER IN PARTS LIST.
3. SEE PARTS LIST FOR FULL COMPONENT DESCRIPTION.

ITEM NO.	DESCRIPTION	QTY
1.	PRINTED CIRCUIT BOARD 400501-1	1
2.	RESISTOR 10K 1/4W 10%	5
3.	1K 1/4W 10%	2
4.	2.2K 1/4W 10%	1
5.	220Ω 1/4W 10%	1
6.	POTENTIOMETER 10K	1
7.	DIODE 1N1662A	1
8.	CAPACITOR .01 MFD 35V 10%	3
9.	CAPACITOR 10 MFD 35V 10%	2
10.	TRANSISTOR 2N3644	4
11.	TRANSISTOR 2N3643	1
12.	FUSE 1/2 AMP	

DRAWN BY:
CHECKED BY:
REVISED:

DENVER INSTRUMENTS CORP.

ASSEMBLY,
PRINTED WIRING BOARD,
20 mA CONTROL LOOP
INTERFACE

SCALE: 4/1   SIZE D: 87-F60461

Fig. 7-33.   An assembly drawing with a parts list. The list is used for parts purchasing and description. The assembly drawing will show manufacturing how and where to "stuff" each component.

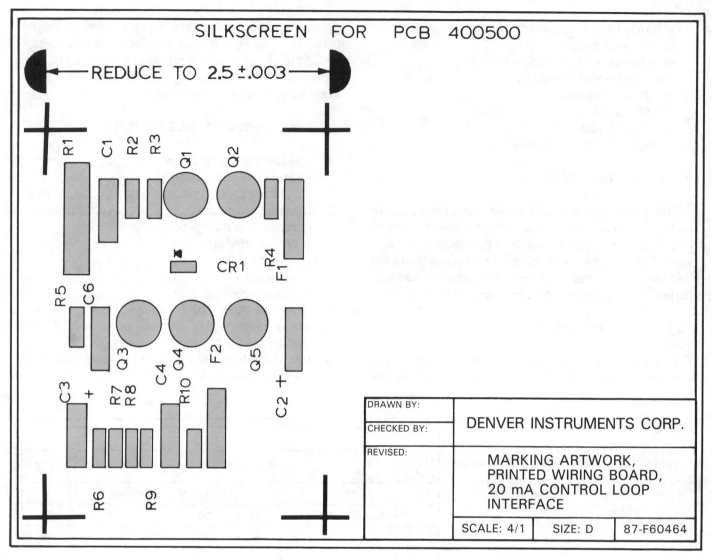

Fig. 7-34. Silkscreen drawing used to show assemblers quickly where each component goes and the polarity of each. Often called marking artwork, it is used to print instructions on the PC board. Note reduction marks.

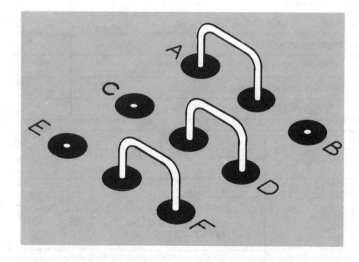

Fig. 7-35. An example of how a silkscreen drawing will assist the assemblers in setting up desired customer's options. Jumper wires re-route the circuit as needed. Instructions may be marked on the PCB.

7. What is the purpose of the drill and trim drawing?
8. An inked layout is _____ a PCB artwork taping.
   a. The result of.
   b. Done before.
   c. The only step before.
9. Why is it necessary to gold plate the PCB edge connector pins?
10. How are photographic targets used in the etching process?
11. Amperage requirements affect what aspect of PCB design?
12. What are the normal scales used when designing a PCB?
13. When would you do a 1 to 1 artwork layout?
14. Why do you use a 10th grid format when laying out a PCB?
15. List the information found on an assembly drawing.

PROBLEMS

PROB. 7-1. Using the BOT-EOT schematic, Fig. 7-36 and components on Fig. 7-37, create a formal schematic. Add reference designations to the schematic.

PROB. 7-2. Design a printed circuit board from the BOT-EOT schematic. The board size should be 2.50 x 4.5 in. Inputs and outputs should be on the same side of the board. The connection terminals should be .375 dia. at 2X scale. It should be a double-sided board, so use a red and blue pencil for the layout.

PROB. 7-3. Tape the BOT-EOT board into a formal artwork. Tape width should be .05 and the spacing .05. Use a 10th grid to control the taping.

PROB. 7-4. Create a drill and trim drawing of the BOT-EOT board.

PROB. 7-5. Make an assembly drawing with parts list of the BOT-EOT board.

PROB. 7-6. Using Figs. 7-38, 7-39, and 7-40, create a printed circuit board layout.

PROB. 7-7. Complete problem 6 by creating an artwork, drill and trim, and an assembly drawing.

PROB. 7-8. Using Figs. 7-41 and 7-42, create a printed circuit board layout, an artwork, drill and trim, and an assembly drawing.

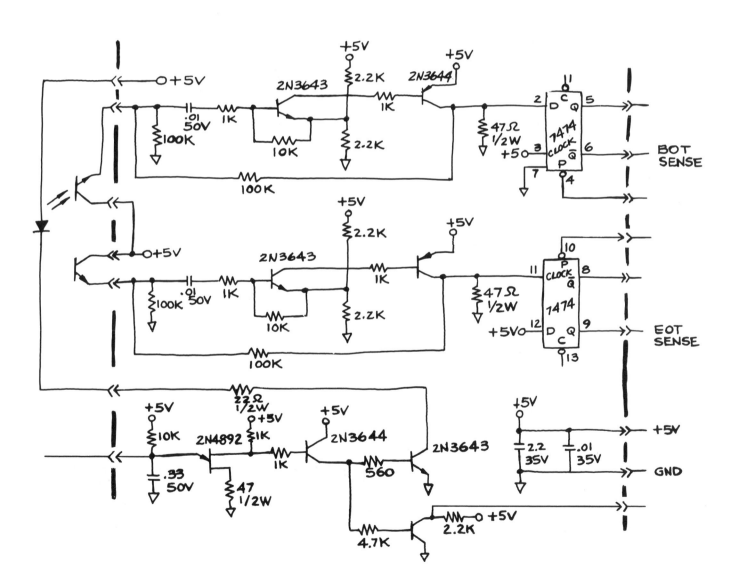

Fig. 7-36. A typical schematic as submitted to drafting by the engineer. Draw a formal schematic.

BOT-EOT BOARD

14 PIN DUAL IN LINE CHIP

RESISTOR ¼ WATT

RESISTOR ½ WATT

TRANSISTOR
2N3643
2N3644

E
B
C

TRANSISTOR
2N4892

1 BASE I
2 EMITTER
3 BASE II

CAPACITOR
.01 UF, 50V
.01 UF, 35V
2.2 UF, 35V

CAPACITOR
.33UF, 50V

- TERMINAL SIZES .24 DIA

- DOUBLE SIDE BOARD

- BOARD SIZE 4.50 × 9.00 @ 2:1 SCALE.

Fig. 7-37. Design information is given for each component for the BOT-EOT board.

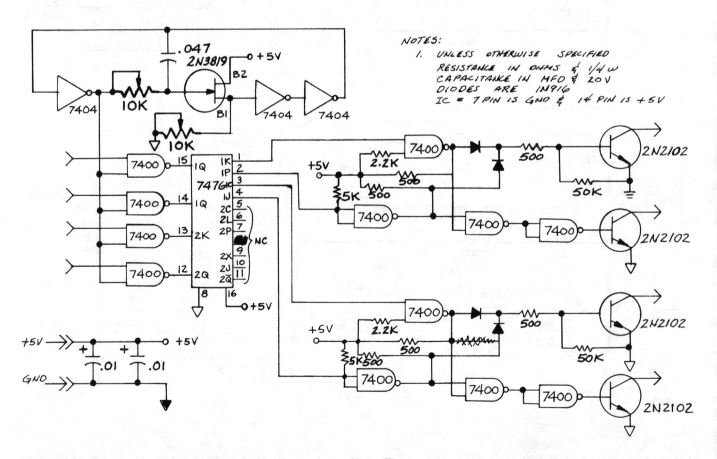

NOTES:
1. UNLESS OTHERWISE SPECIFIED RESISTANCE IN OHMS & ¼ w CAPACITANCE IN MFD & 20 V DIODES ARE 1N916 IC = 7 PIN IS GND & 14 PIN IS +5V

Fig. 7-38. A typical logic drawing supplied by the engineer. Note: There are some areas marked out and one ground symbol is incorrect. These are errors the drafter cleans up. Create a printed circuit board layout from this and two other drawings.

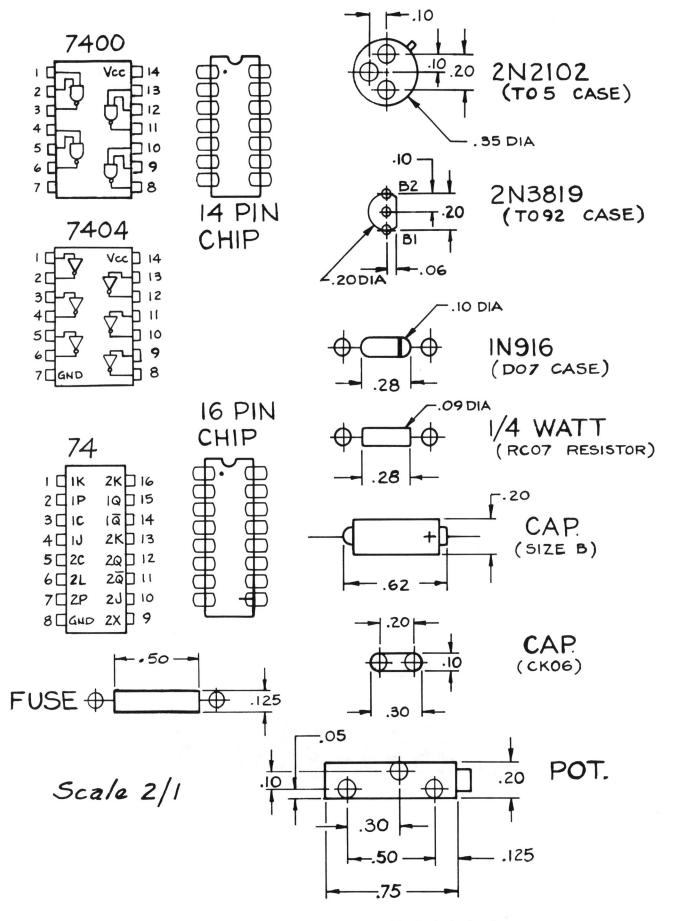

Fig. 7-39. Component styles for Fig. 7-38, the logic drawing.

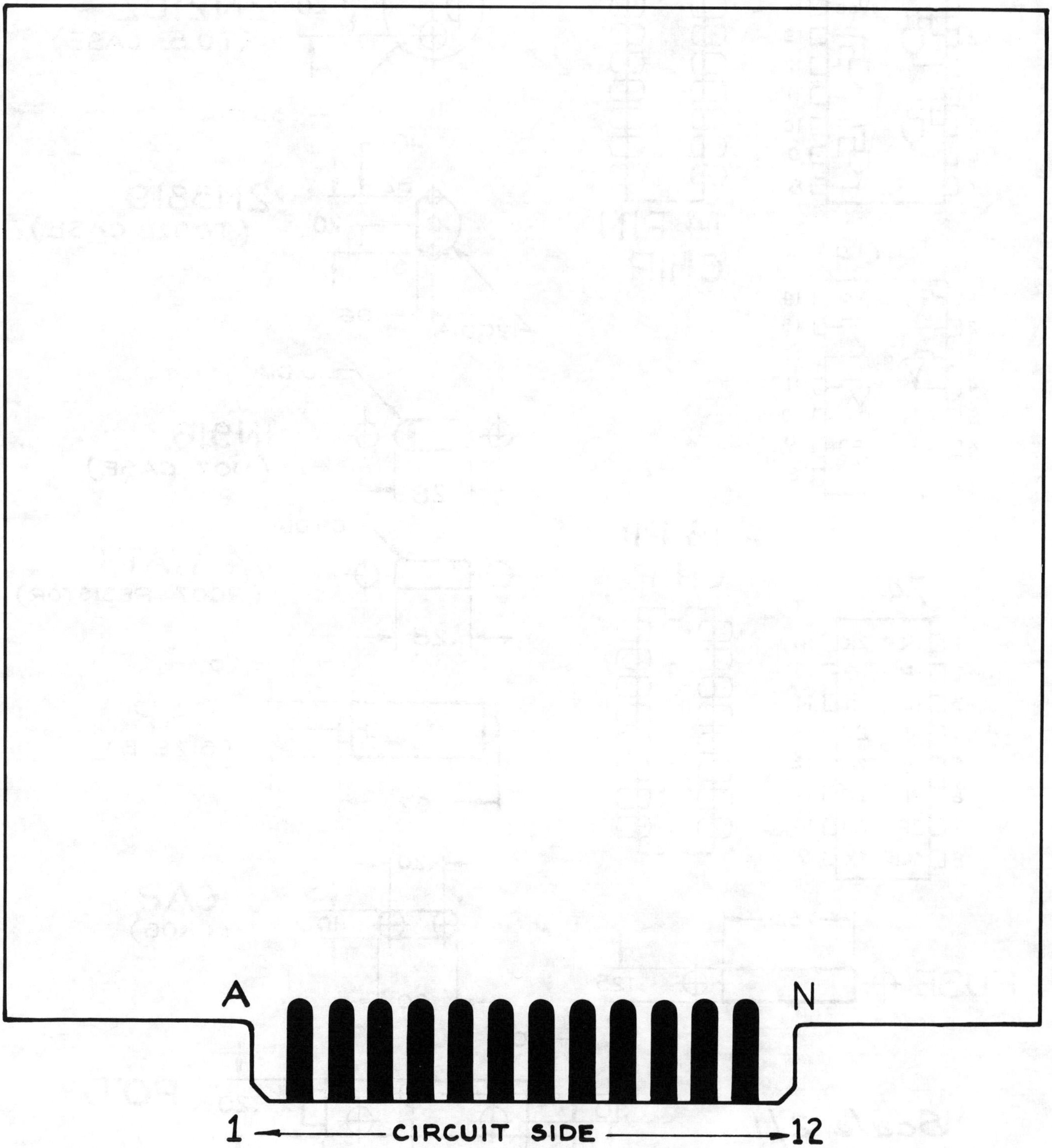

A　　　　　　　　　　　　　　　　　　　　　　N

1 ← CIRCUIT SIDE → 12

SCALE 2/1

Fig. 7-40. Board size and connector for Fig. 7-38.

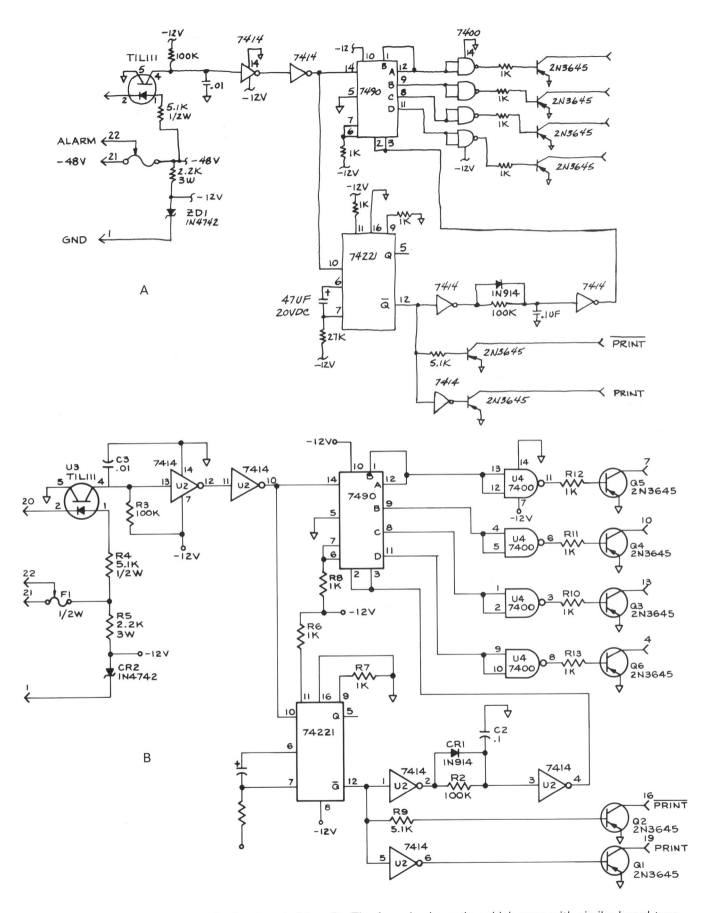

Fig. 7-41. A—Schematic as submitted to drafting. B—The formal schematic, which goes with similar board type not using "I, O, Q" when counting A to N. First, create a formal schematic, then a PCB layout from the schematic and next drawing.

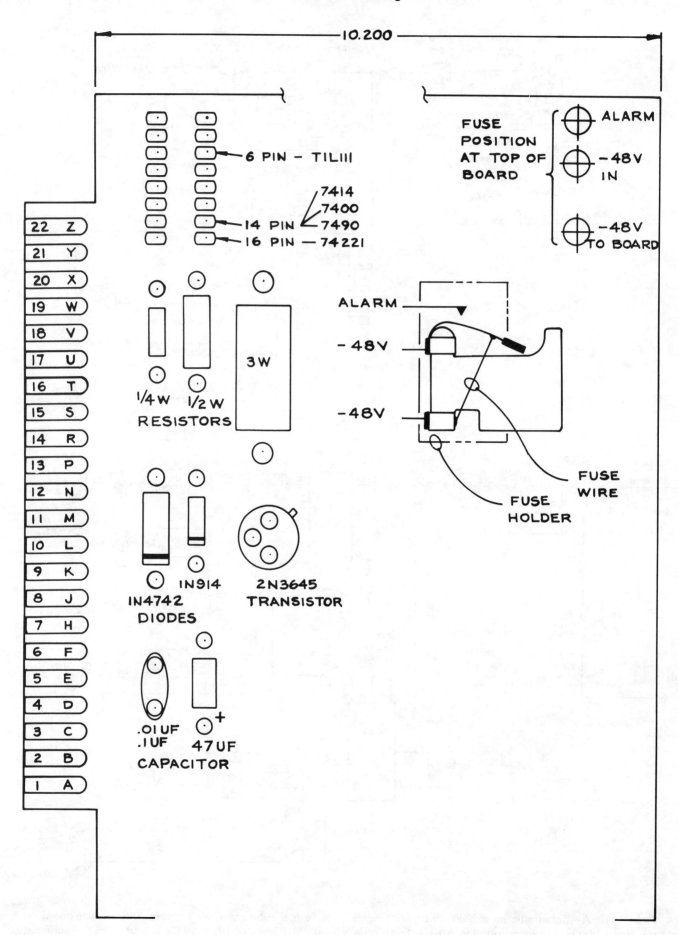

10.200

6 PIN – TIL111

7414
7400
14 PIN 7490
16 PIN – 74221

FUSE POSITION AT TOP OF BOARD

ALARM
–48V IN
–48V TO BOARD

22 Z
21 Y
20 X
19 W
18 V
17 U
16 T
15 S
14 R
13 P
12 N
11 M
10 L
9 K
8 J
7 H
6 F
5 E
4 D
3 C
2 B
1 A

3W

1/4 W 1/2 W
RESISTORS

IN914
IN4742
DIODES

2N3645
TRANSISTOR

.01 UF
.1 UF
47 UF
CAPACITOR

ALARM
–48V
–48V

FUSE WIRE

FUSE HOLDER

Fig. 7-42. Components, board size, and connector data for Fig. 7-41.

Chapter 8

PACKAGING DRAWINGS

After studying this chapter you will be able to:
- State the process and identify fasteners for fastening metal parts.
- List the qualities of electronics packaging materials.
- Describe welding processes and apply symbols to call out welds.
- Calculate bend allowances for sheet metal packaging and lay out stretchouts for sheet metal parts.
- Apply rules of good dimensioning to mechanical drawings.

INTRODUCTION

The word packaging refers to the process of putting the electronics into a housing designed for the environment and for safety. The career title for this area of drafting work is Packaging Drafter. The field involves drafting of structural shapes for solid parts, rather than drafting of electronic circuits. Electronics drafting is involved only as a guide to part locations, mounting methods, and other mechanical requirements.

In this chapter, you will study the processes needed to make these enclosures. You will start your study with fastening processes.

FASTENING

The basic means of putting parts together is called fastening. Fastening is accomplished by welding, mechanical fasteners, or bonding with adhesives. These fastening methods are used at the subassembly and assembly level.

WELDING

Welding can be used wherever we want the parts permanently joined. Welding is accomplished by using heat, pressure, or both. The preferred welding methods are: gas, arc, resistance, and adhesive. Fig. 8-1 shows the process of arc welding.

Gas welding makes use of burning gases to produce the welding temperature, Fig. 8-2. When the heat melts the metals, they will fuse together. Filler material may also be used in this process. The filler material will normally be similar to the composition of the base metals.

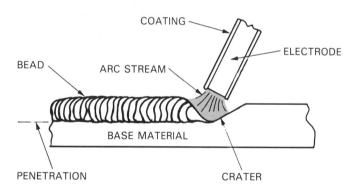

Fig. 8-1. Arc welding is a joining process that uses an electric arc to produce the heat required to melt the metals. The melting of the metals allows them to fuse together. Filler metal from the electrode may be added to the joint.

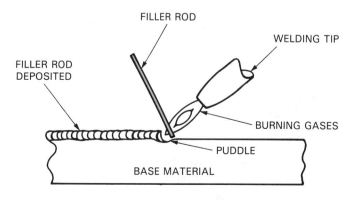

Fig. 8-2. Gas welding using a filler rod to deposit additional material.

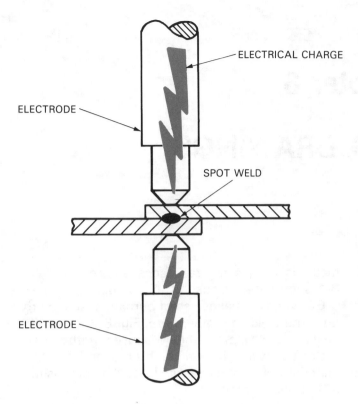

Fig. 8-3. Materials to be spot welded are placed between electrodes. The electrodes are then pressed together while an electrical charge is applied. This charge melts a spot between the metals because of their resistance. When the spot has cooled, the electrodes open up. This allows the part to be moved to the next spot weld or removed for the next part.

Resistance welding is the process of welding by heat and pressure. Spot welding is the most common form of resistance welding, Fig. 8-3. The weld is produced by heat obtained from the resistance of the material to the flow of electrical current. This process is used for its advantages in electronics packaging. Resistance welding is economical and gives weight savings, since it uses no additional filler metal. The welds are made directly between the metal parts being joined.

MECHANICAL FASTENERS

Mechanical fasteners have been used for centuries to join parts together. In the family of mechanical fasteners there are screws, bolts, pins, and keys. These fasteners allow the joined parts to be easily separated. Riveted and press-fit joints are considered more permanent.

Fastening parts with mechanical fasteners requires skill in the preparation of the parts being joined. It is a costly way to join parts, but maintenance and inspection of some parts require this method for ease of disassembling.

SCREWS

Screws use the principle of the wedge to clamp an assembly together. Screws are the most common mechanical fastener in electronics.

Machine screws are available in many different head styles, Fig. 8-4. Each screw type has a special application. For example, if you want a screw to be flush with the material being joined, you would select a flat head screw. These screws can be turned by different drivers, Fig. 8-5. The socket and hex head screws allow a fastener to be tightened to a high torque level.

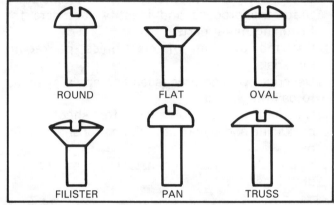

Fig. 8-4. Some of the head styles of machine screws.

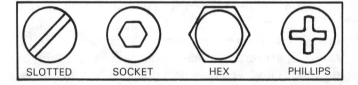

Fig. 8-5. Different screw drives are required for each of these head types.

Screw Threads

Machine screws come in three main thread types, unified national coarse (UNC), fine (UNF), and extra fine (UNEF). The number of threads per inch in relationship to the diameter dictates the thread type, as listed in Fig. 8-6.

SELF-TAPPING SCREWS

In electronics, there are many applications for self-tapping screws. Refer to Fig. 8-7. Self-tapping screws work well on thin material and save the expense of the costly tapping operation. This thread type also works well in plastic parts.

MACHINE SCREWS				
Size	Diameter	Coarse UNC	Fine UNF	Extra Fine UNEF
0	.060	—	80	
1	.073	64	72	
2	.086	56	64	
4	.112	40	48	
6	.138	32	40	
8	.164	32	36	
10	.190	24	32	
12	.216	24	28	
1/4	.250	20	28	
5/16	.3125	18	24	
3/8	.375	16	24	32
7/16	.4375	14	20	28
1/2	.500	13	20	28

Fig. 8-6. A table of machine screw sizes including thread classification by threads per inch.

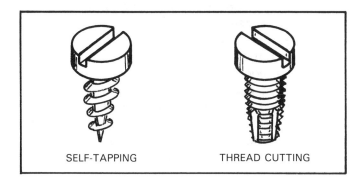

SELF-TAPPING · THREAD CUTTING

Fig. 8-7. Sheet metal screws include a self-tapping screw and a thread cutting screw that requires a clearance hole to start the thread.

BOLTS

Machine bolts are used to assemble items that do not require close tolerance fasteners. They are manufactured with square, hex, and socket heads, Fig. 8-8. Bolts are normally furnished with nuts.

PINS

Pins are used to retain parts in position. See Fig. 8-9. With a washer, pins may be fitted into a hole drilled crosswise in a shaft to prevent parts from slipping or spinning off.

WASHERS

Washers provide an increased surface area for bolt heads and nuts. They distribute the load over a larger area. There are four main washers used in electronics: flat washers, split-ring lock washers,

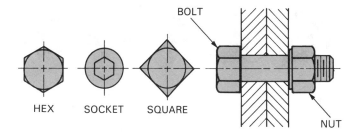

HEX · SOCKET · SQUARE · BOLT · NUT

Fig. 8-8. A bolt and nut with examples of head types.

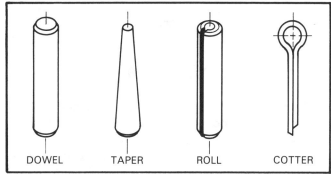

DOWEL · TAPER · ROLL · COTTER

Fig. 8-9. Pins are used for locating and holding parts in position.

external-tooth lock, and internal-tooth lock. Examples are shown in Fig. 8-10.

Flat washers come in four different styles: light, medium, heavy-duty, and extra heavy-duty. Flat washers prevent marring of the surface finish by turning. Split-ring lock washers are used to prevent fasteners from loosening under vibration. External-tooth lock washers have electronic advantages other than their superior holding power. They cut into the chassis when the fastener is tightened, making it possible to get a high quality ground connection. Internal-tooth lock washers are not as effective as external type but they are desirable when appearance dictates a clean non-snagging surface.

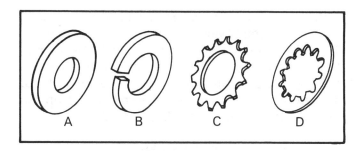

A · B · C · D

Fig. 8-10. The most often used washers in the electronics industry. A—Plain flat washer. B—Split-ring type. C—External-tooth lock washer. D—Internal-tooth lock washer.

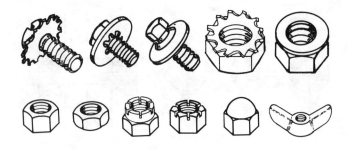

Fig. 8-11. A—Some screws and nuts are pre-assembled with washers. There are many nut styles. B—Standard nut. C—Jam nut. D—Slotted and castellated nuts. E—Cap nut. F—Wing nut.

PRE-ASSEMBLED FASTENERS

Pre-assembled fasteners help cut assembly time. When assembling areas are confined, it is hard to handle separate nuts and washers and secure them to a bolt. Therefore, pre-assembled fasteners save costs and reduce assembly time. Fig. 8-11A shows some screw units and nut units.

NUTS

Nuts normally have an external hexagonal or square head. They are used with bolts or studs. They come in different types and some have special applications.

The standard nut is the type most used, Fig. 8-11B. Other nut types are common. The jam nut, Fig. 8-11C, is a thinner nut often used to lock a standard nut in place. Castellated and slotted nuts, Fig. 8-11D, are manufactured so that you can use a cotter pin to prevent them from loosening. Cap nuts, Fig. 8-11E, are applied when appearance and safety are important. Wing nuts, Fig. 8-11F, can be loosened and tightened without the use of a wrench. Its shape allows hand adjustment.

RIVETS

Rivets are used to make permanent assemblies. They have been developed for different applications. Therefore, they have different head styles, Fig. 8-12. A rivet type used frequently in electronics assembly is the pop rivet, or blind rivet, Fig. 8-13. It has the benefit of not needing a back-up on the far side of the material being riveted.

ADHESIVES

Adhesives are often used to create permanently bonded joints. They can be applied to almost any

material. The new synthetic adhesives like epoxy-resin based adhesives are used for many electronic packaging applications. They give a clean appearance and make strong joints.

CLEANING SURFACES

The finishing process cleans, protects, and decorates the surface. Cleaning the surface to free it of dirt, oils, and rust is usually accomplished to prepare the metal for further treatment. Cleaning can be accomplished by chemical or mechanical means. Mechanical methods use abrasives, sand blasting, scraping, and wire brushing. Chemical methods use alkaline cleaning. Once the metal has been cleaned, we must select the appropriate finish.

FINISHES

Finishes are applied for different purposes. Some of these purposes are:
1. Protective coatings: coatings applied to the metal to protect it from corrosion and abrasion, and to hide imperfections.
2. Chemical coating: a thin layer of metallic compound produced by chemical or electrochemical treatment of the surface. Types of finishes are chromates, oxides, and anodized films.
3. Organic coating: a layer of paint, lacquer, enamel, varnish, or primers sprayed or brushed on the metal.

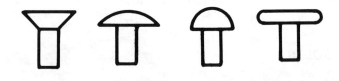

Fig. 8-12. Rivet styles. A—Countersink. B—Truss head. C—Round head. D—Flat head.

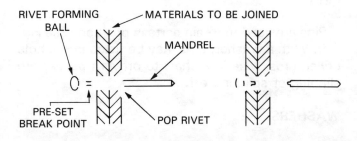

Fig. 8-13. A—Rivet being set. B—The forming ball has been pulled into the rivet, forming a head. The mandrel is preset to break after the forming ball has been seated. The mandrel is pulled with a pop-rivet tool.

4. Metallic coating: a film of metal or metal alloy deposited by chemical, electroplating, hot dipping, or electrodeless plating processes.
5. Lubricant coating: a film applied to the metal to reduce surface friction. Types of films are oils, waxes, and mineral substances such as graphite.

METALS AND FINISHES

For each metal you must select an appropriate finish. Metals predominately used in electronics packaging are aluminums and steels. The following is a breakdown of standard finishes for these metals.

Aluminums are generally anodized, but they may also be caustic etched or phosphate coated. These preparations can be a preliminary step to the application of organic coatings. Zinc chromate undercoatings will also prepare aluminum for painting.

Steels can be plated with cadmium, copper, nickel, chromium, silver, and gold. An electronic chassis is normally cadmium plated. Zinc chromate undercoatings will prepare steel for organic coatings.

MATERIALS FOR PACKAGING

Material selection for the packaging of electronics components normally involves three areas: plastics, steels, and aluminum material. We begin our study with plastics.

PLASTICS

Plastics are being used more as electronics packages become smaller. Plastics are easily formed into intricate shapes, Fig. 8-14. Plastics can also be welded or machined. Reinforcing materials can be added to plastics to make them more rigid and to give them strength required in some applications.

There are two general classifications of plastics: thermosetting and thermoplastic. Thermoplastics are those materials which can be heated and reformed. They can also be remelted and recast like metals. In contrast, thermosetting plastics cannot be reformed once they are solidified into their permanent shape.

THERMOSETTING

Thermosetting plastics are classified into groups. Epoxy, urethane, phenolic, and melamine are some of the group classifications. Epoxy plastics are used in electronics as adhesives, and for electrical insulation, structures resistant to acids, and coating applications. Urethanes are used for good thermal and electrical insulation. Phenolics have hard and elastic properties. They are used for electronics cabinets and printed circuit board applications. Melamines are hard and strong. They can be easily colored and have good resistance to oils. They are used for electrical wiring devices such as terminal blocks.

THERMOPLASTICS

Thermoplastic plastics come in the following groups: ABS (acrylonitrile-butadiene, styrene), acrylics, nylon, vinyls, and cellulosics. ABS plastics are strong and weather resistant. They are used in telephone components. Acrylics are strong and rigid. Their electronic applications are panel windows, light covers, and cabinet paints. Nylons are

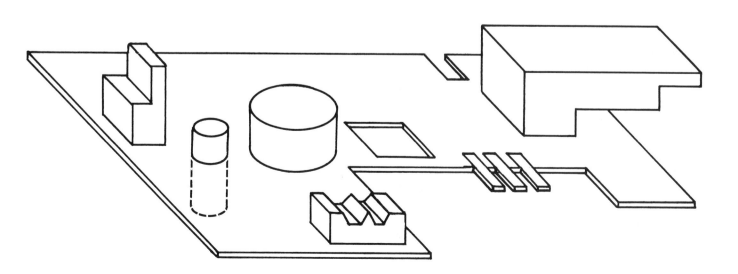

Fig. 8-14. Support structure for computer disk drive.

tough and elastic. They are used for bushings, gears, washers, and hinges. Vinyls are used to add decorative surfaces to electronic cabinets. Cellulosics are tough and have good insulating properties. They are used in the manufacture of drafting templates and equipment. A common electronic use is in telephone sets.

FABRICATING WITH PLASTICS

Plastics can be purchased in standard forms, such as sheets, tubes, structural shapes, and laminates. They can also be purchased as raw material in powders, liquids, or granules.

Plastics bought in standard forms can be cut into shapes, machined to drawing specifications, or heated and reshaped to desired forms. Thermoforming plastics can be heated and blow-molded or vacuum-formed, Fig. 8-15. Thermoforming has limitations in what shapes can be produced. When you want intricate shapes, they can more easily be injection molded.

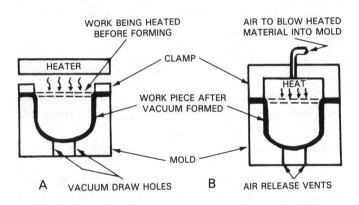

Fig. 8-15. A—In vacuum-forming, a plastic sheet is heated and drawn into a mold with a vacuum. B—A blow-forming process.

Injection molding normally requires thermoplastics. The plastic will be purchased in a powder or granular form. It is heated in an injection hopper. When it is in a plastic state, it will be forced by a plunger into a mold, Fig. 8-16. Another way to form plastics is by an extrusion process. Extrusions are made by forcing semiliquid plastic through a preshaped die, Fig. 8-17. Unlimited numbers of shapes can be made by this process. The die can be changed to allow a new shape to be formed.

ALUMINUMS

Aluminums have at least seven properties which make them desirable for use in electronics. Here is

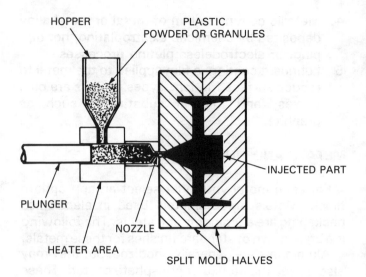

Fig. 8-16. A simplistic look at the plastic injection process. Note how the plunger forces the molten plastic through the nozzle and into the mold cavity.

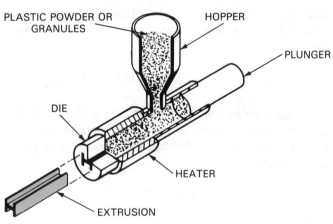

Fig. 8-17. An illustration of a plastic extrusion process.

a list of properties:
1. Weight—aluminum weighs less than other metals.
2. Ease of fabrication— it can be easily formed or machined.
3. Corrosion resistance—it needs no protection in most ordinary environments.
4. Finishes easily—there are many finishes which can be economically applied to aluminum.
5. Electrical properties—pound for pound it is twice as conductive as copper. However, with the same cross-sectional area, it conducts 62% as well as copper.
6. Nonmagnetic—this quality makes it a good shield in electronic cables and equipment.
7. Heat dissipation—it transmits heat rapidly. This makes it useful for heat-sinking electrical components that would otherwise burn out.

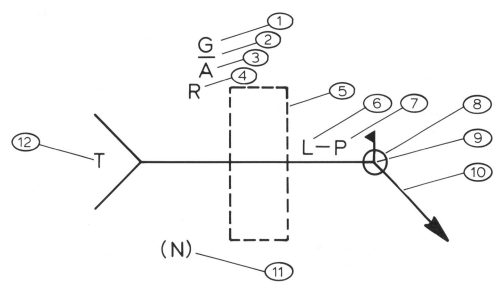

1. SPECIFIES THE FINISH ON THE WELD. G = GRIND, C = CHIP, M = MACHINE.
2. MEANS THE WELD WILL BE FLUSH WITH THE SURFACE. ⌒ MEANS A CONVEX SHAPED WELD.
3. SPECIFIES THE INCLUDED ANGLE OF THE COUNTERSINK FOR A PLUG WELD.
4. SPECIFIES THE ROOT SIZE OF A PLUG OR SLOT WELD, FIG. 8-19.
5. LOCATION OF THE BASIC WELD SYMBOL, FIG. 8-20.
6. LENGTH OF WELD IN INCHES, FIG. 8-21.
7. PITCH OR CENTER TO CENTER SPACING OF NONCONTINUOUS WELDS, FIG. 8-21.
8. SPECIFIES WELD AROUND COMPLETE JOINT, FIG. 8-22.
9. SPECIFIES WELD IN FIELD. THIS IS USED WHEN ONLY SOME OF THE WELDING WILL BE COMPLETED IN THE SHOP.
10. ARROW POINTING TO WELD JOINT OR GROOVED MEMBER, OR TO BOTH.
11. SPECIFIES NUMBER OF SPOT WELDS, FIG. 8-23.
12. A CODE LETTER USED TO REFERENCE A GENERAL NOTE EXPLAINING WELDING SPECIFICS. IF THERE IS NO CODE LETTER USED THE ENTIRE TAIL SHOULD BE OMITTED.

Fig. 8-18. An American Standard Graphic Symbol for welding. The standard it follows is ANSI/AWS A2.0. See table above for explanation.

STEELS

Steels are playing a decreasing role in electronics packaging. Small electronic components do not require the strong construction of steel packages.

Carbon steels are the types most used in industry. They make up 85% of all steels produced. Carbon steels are made by hot or cold rolling them into sheets or shapes. Hot-rolled steels have the lowest cost per pound but they lack some qualities of cold-rolled steels. Cold-rolled steels (CRS) are most often used in electronics.

DRAWINGS FOR WELDING

In order to adequately specify welding requirements, you must first see how the symbols are used to represent the required joint and weld. First, you will study the welding symbol, Fig. 8-18. There are also weld symbols for the types of welds used, Figs. 8-19 through 8-24. A typical welding drawing is shown in Fig. 8-25.

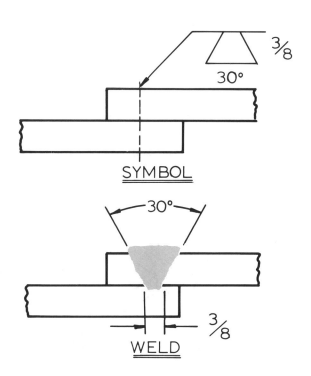

Fig. 8-19. How to identify a plug or slot weld.

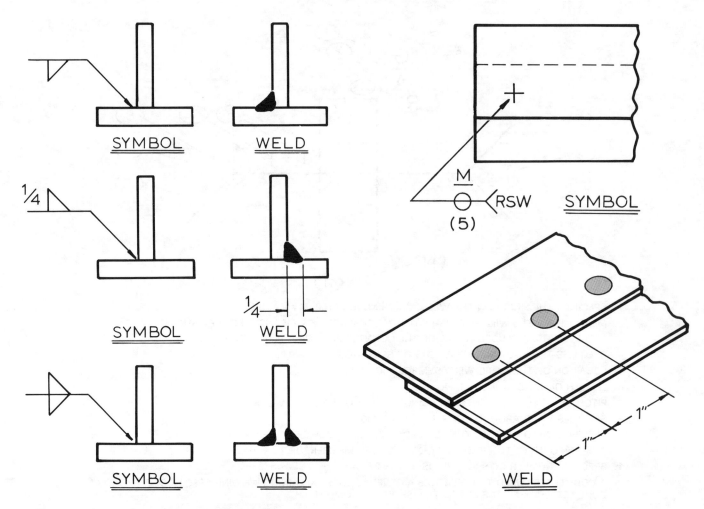

Fig. 8-20. How to indicate on which side or sides a weld is placed.

Fig. 8-23. Symbol means five spot welds located 1 in. apart and machined flat.

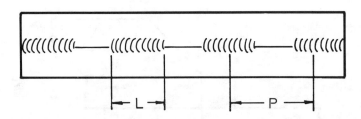

Fig. 8-21. What the length and pitch mean on a non-continuous weld.

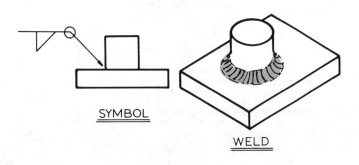

Fig. 8-22. Example of symbol to weld all around the joint.

WORKING WITH SHEET METAL

Electronics drafters design many sheet metal parts. The drafter must understand sheet metal and what happens to it when it is bent or formed.

Sheet metal parts are cut from flat stock using patterns or templates created by drafters. After pieces are cut out of the flat stock, they are usually formed by a BRAKE, Fig. 8-26. A brake will normally create straight lines and non-complex parts.

Complex parts will be made from metal blanks. The blanks will be punched or pressed over or into dies, Fig. 8-27. These parts will normally need to be trimmed after being formed. See Figs. 8-28 and 8-29 for sheet metal drawings. Fig. 8-28 shows the part in a flat state. Fig. 8-29 shows the part after it has been developed.

BENDING SHEET METAL

When sheet metal is bent, there is a local thinning or thickening of the material. This is caused

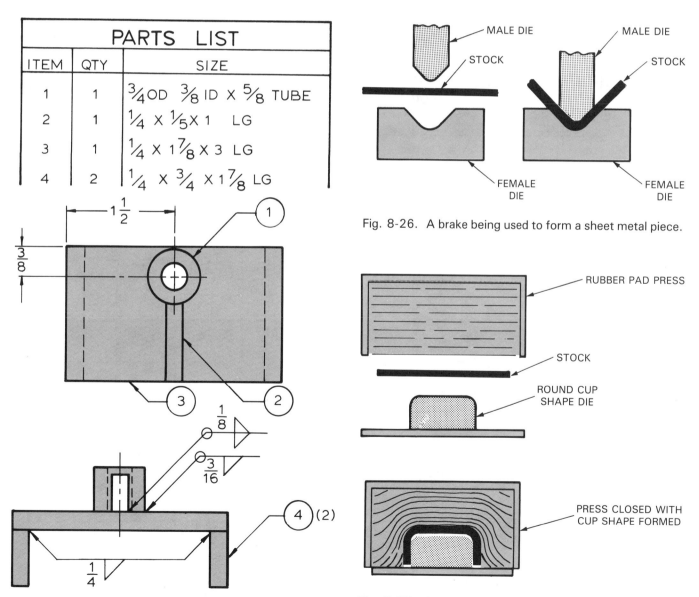

	GROOVE						FLANGE	
SQUARE	V	BEVEL	U	J	FLARE-V	FLARE-BEVEL	EDGE	CORNER
‖	∨	∨	∪	⊍	∨	⊬	⌐	⫟

FIELD WELD	WELD ALL AROUND	FILLET	SPOT PROJECTION	PLUG OR SLOT	SEAM	BACK OR BACKING	SURFACING
▶	⊙	◣	○	▭	⊖	⌣	⌣⌣

Fig. 8-24. Symbols used to indicate the type of preparation and weld used in the joining process.

PARTS LIST

ITEM	QTY	SIZE
1	1	$\frac{3}{4}$ OD $\frac{3}{8}$ ID X $\frac{5}{8}$ TUBE
2	1	$\frac{1}{4}$ X $\frac{1}{5}$ X 1 LG
3	1	$\frac{1}{4}$ X $1\frac{7}{8}$ X 3 LG
4	2	$\frac{1}{4}$ X $\frac{3}{4}$ X $1\frac{7}{8}$ LG

Fig. 8-25. A typical welding drawing.

Fig. 8-26. A brake being used to form a sheet metal piece.

Fig. 8-27. A rubber pad is used to force the metal blank down smoothly around the cup-shaped die.

113

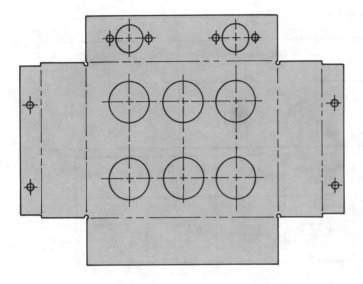

Fig. 8-28. A flat piece of sheet metal trimmed to size and punched with appropriate holes.

by the metal stretching on the outside of the bend and compressing on the inside. This causes plastic deformation of the material. Through studying many sheet metal parts, it has been found that there is a line through the bend where no stretching or compression takes place. This line is called the MEDIAN LINE, and the line is used for bend allowance calculations, Fig. 8-30.

SHEET METAL TERMS

In order to fully understand the following information, you must first familiarize yourself with some basic terms:

Bend allowance: The length of material around a bend. Length is calculated on the median line, as indicated in Fig. 8-30.

Bend angle: The angle through which sheet metal

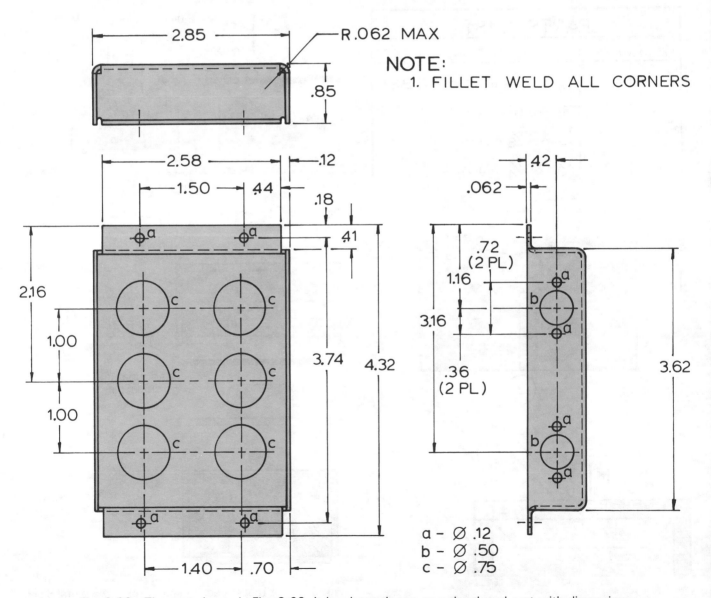

Fig. 8-29. The part shown in Fig. 8-28. It has been drawn as a developed part with dimensions.

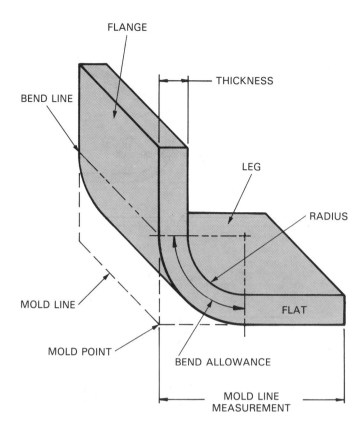

Fig. 8-30. Sheet metal terms and what they mean.

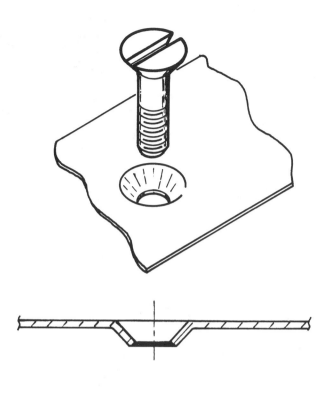

Fig. 8-31. A thin section of sheet metal dimpled to fit a screw head.

is bent. The angle is measured on the outside of the bend. It is measured from the flat around the bend to the finished angle.

Bend line: The line formed by the beginning of the bend and the end of the bend. It is the tangent point on both intersections of the bend and flat surface.

Center line of bend: A radial line from the center of the bend radius and bisecting the included angle between bend lines.

Developed length: The length of a flat sheet which can be bent to create a part. This length is shorter than the sum of the mold line dimensions on the part.

Dimensioning and tolerancing: Dimensioning sheet metal parts formed on a brake is referred to the outside mold lines and inside bend radii. Dimensioning to inside mold lines will be done only if the inside finished dimensions are critical. *If the inside dimensions are critical, the part should be produced by a die.* Normally you will not tolerance brake-formed parts closer than ±.03 for bend radii or between bends. See Fig. 8-33.

Dimpling: Stretching a relatively small shallow indentation in the sheet metal. It can be used for countersinking for screw heads, Fig. 8-31.

Flat pattern: A flat layout of a sheet metal piece to be formed. The metal can be bent to make the

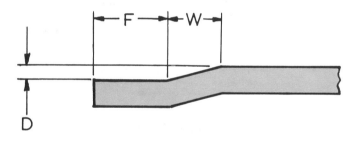

Fig. 8-32. A joggle shape with standard dimensioning technique.

finished part without trimming.

Joggle: An offset in the sheet metal. It is used to give strength, design, or to set the edge for a mating part, Fig. 8-32.

Lightening holes: Holes or cutouts in sheet metal used to reduce weight. These holes are not functionally critical so they should be dimensioned for liberal sizes.

Mold line: The imaginary line of intersection of the two flat surfaces of the formed sheet metal part, as in Fig. 8-30.

Set-back: The amount of deduction in length of the flat pattern. This deduction is the saving in material by going around a bend radius rather than around a square or sharp corner. See Fig. 8-30.

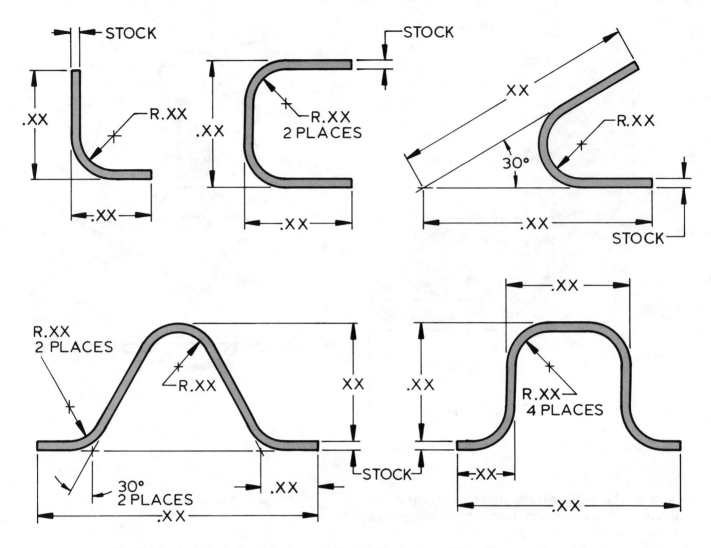

Fig. 8-33. Standard dimensioning techniques for sheet metal profile or section views.

CALCULATING BEND ALLOWANCE

To determine the distance around a bend, we must know three things. Find out the following:
1. The radius we will bend around. This will normally be determined by a bending table. See appendix.
2. Thickness of the material.
3. The angle of the bend.

If the radius is one inch or less, we can use this formula for bend allowance (BA):

BA = (.0078T + .01745R) × Degree of bend.
T = Material Thickness = .125
R = Inside Bend Radius = R.25

To see how this applies, see Fig. 8-34. For the example in Fig. 8-34, BA equals 0.800.

BEND DIMENSIONING

Dimensioning of a bend will be done by establishing the mold point or mold line. To establish the mold lines, we can use trigonometry. The formula is:

$$X = (R + T) \div (TAN \frac{\alpha}{2})$$

X = distance from bend line to mold line

R = radius (inside)

T = thickness of material

α = included angle between flanges

TAN = TANGENT

Using Fig. 8-35, calculate the dimensioning

R = .25, T = .125, α = 30°

$$X = (.25 + .125) \div (TAN \frac{30°}{2})$$

$$\div TAN 15$$

$$.375 \div .268$$

X = 1.400

This number is used in the next calculation step.

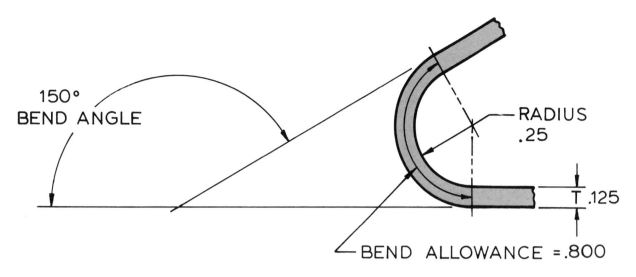

Fig. 8-34. Necessary information for figuring bend allowance. BA = 0.800 in this example.

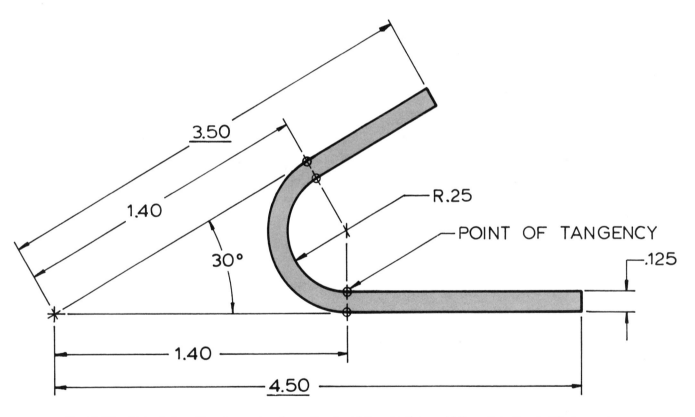

Fig. 8-35. Calculating the dimension from the bend line to the mold line. It is the 1.40 dimension.

FIGURING SET-BACK FOR FLAT PATTERN

There is a formula to figure the deduction made in the length of a flat pattern. The deduction is equal to twice the distance from the bend line to the mold line minus the bend allowance. Look at the formula:
SET-BACK (K) = (2X − BA)

X = mold line to tangent point dimension
BA = bend allowance

Using the information from the previous two figures and this formula, you can design a flat pattern:

K = 2 (1.400) − .800
K = 2.800 − .800
K = 2.000

See Fig. 8-36 for the flat pattern drawing. With a perfectly sharp bend, the full length would be equal to 4.50 plus 3.50. But it is really 4.50 plus only 1.50 because the set-back of 2.00 was subtracted.

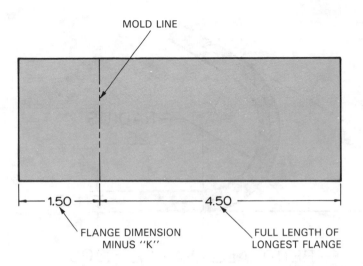

Fig. 8-36. A flat pattern dimensioned after figuring the set-back distance. The set-back, called "K," equals 2.00.

CORNER BEND RELIEF

Whenever two sheet metal bends intersect one another, they require a corner relief cutout. Without this relief, the corner would buckle or wrinkle. Material must be removed .03 inches minimum behind bend lines, Fig. 8-37.

LIMITATIONS IN SHEET METAL

In order to bend a flange on a piece of sheet metal, we need a specific amount of material. Minimum widths are determined by material thickness and bend radius. Minimum distances between bends must also be allowed. A power brake can only work to set limits. The limits can be determined by adding the minimum flange width from the table in Fig.

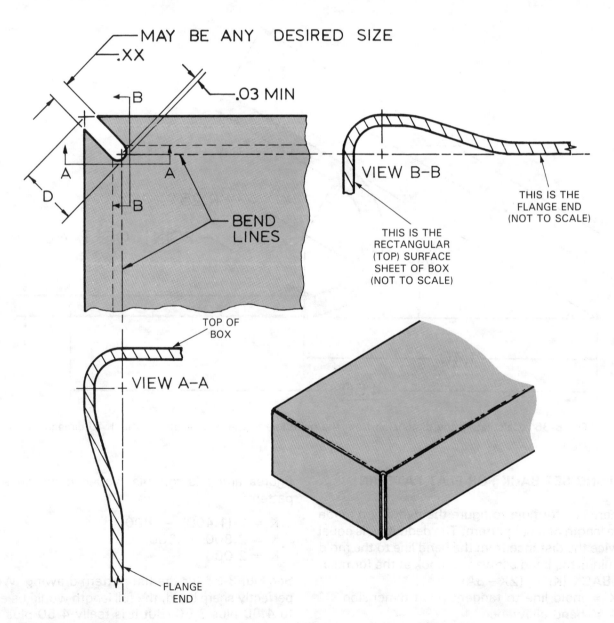

Fig. 8-37. Corner bend relief is required when two bends intersect. Here the cup shape faces away from you.

GAGE (in.)	BEND RADIUS, (in.)												
	.03	.06	.09	.12	.16	.19	.22	.25	.28	.31	.38	.44	.50
.016	.080	.141	.203	.266	.328	.391	.453	.516	.578	.641			
.020	.091	.145	.207	.270	.332	.395	.457	.520	.582	.645	.770		
.025	.107	.150	.212	.275	.337	.400	.462	.525	.587	.651	.775	.900	1.025
.032	.128	.158	.219	.282	.344	.407	.469	.532	.594	.657	.782	.907	1.032
.040	.152	.182	.227	.290	.352	.415	.477	.540	.602	.665	.790	.915	1.040
.051		.215	.246	.301	.363	.426	.488	.551	.613	.676	.801	.926	1.051
.064			.285	.316	.376	.439	.501	.564	.626	.689	.814	.939	1.064
.072			.309	.340	.384	.447	.509	.572	.634	.697	.822	.947	1.072
.081				.367	.398	.456	.518	.581	.643	.706	.831	.957	1.081
.091				.397	.428	.466	.528	.591	.653	.716	.841	.966	1.091
.102					.462	.493	.539	.602	.664	.727	.842	.977	1.102
.125						.562	.593	.625	.687	.750	.875	1.000	1.125
.156								.717	.749	.781	.906	1.031	1.156
.188										.876	.938	1.063	1.188
.250											1.125	1.188	1.250

MINIMUM FLANGE WIDTH
Formed on Power Brake

Fig. 8-38. Use this table to figure minimum flange widths.

8-38 to the bend radius plus material thickness.
To calculate, use the following:
T (Thickness) = .051
R (Bend Radius) = .25
W (Minimum) (From Table) = .55
L (Minimum Flange Length)
FORMULA
L = W + R + T
L = .55 + .25 + .05
L = .85

The minimum flange length is .85 in this example. This means a bend can be no closer than .85 in. to any edge of the flat pattern. It also means bends can be no closer than 2L. Here, 2L equals 1.70 in.

With this information, you can now consider dimensioning sheet metal drawings.

DIMENSIONING

Dimensioning skill is the main difference between drafters and designers. In order to be a designer, you must understand the form, fit, and function of the part or parts. With this understanding, you can dimension the part correctly. Poor dimensioning can add cost to a part whereas good practices will provide a quality part at low cost.

METHODS OF DIMENSIONING

The method you use may not be your main concern, but being sure that you convey the correct information to the machinist is important. Fig. 8-39 shows three dimensioning methods for dimensioning the same part. Each of these methods can control the part, but industry prefers the tabular method on complicated parts.

The tabular method requires less time because you do not have to figure where to place all the extension, dimension, and leader lines. It also sets up the part for numerically controlled machining. This part will be manufactured using the X-axis and Y-axis as datum lines. Each one of the three methods shown uses the same datum corner. Datum selection should be based on the functional requirements of the part. Second, you should consider the manufacturing and inspection of the item. The datum must be easily recognizable and accessible so that all measurements from it can be easily made. Mating parts should use the same datum.

The above paragraphs may seem like the requirements for a mechanical drafter. They are requirements for both mechanical and electronics drafters.

Mechanical drafting knowledge is imperative for

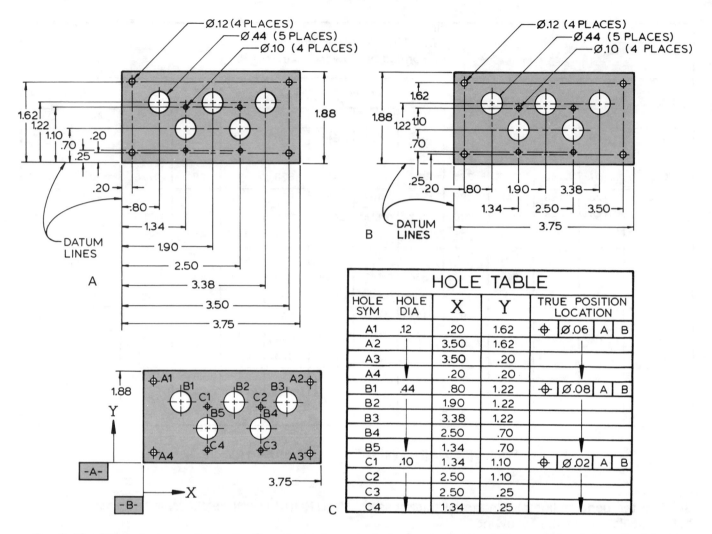

Fig. 8-39. Dimensioning methods. A—Coordinate dimensioning using the bottom left corner as a datum. B—An example of a coordinate system where one arrowhead is used on the dimension line. C—Tabular dimensioning. The last column gives tolerance referred to surfaces A and B.

HOLE TABLE						
HOLE SYM	HOLE DIA	X	Y	TRUE POSITION LOCATION		
A1	.12	.20	1.62	⊕	⌀.06 A	B
A2		3.50	1.62			
A3		3.50	.20			
A4		.20	.20			
B1	.44	.80	1.22	⊕	⌀.08 A	B
B2		1.90	1.22			
B3		3.38	1.22			
B4		2.50	.70			
B5		1.34	.70			
C1	.10	1.34	1.10	⊕	⌀.02 A	B
C2		2.50	1.10			
C3		2.50	.25			
C4		1.34	.25			

electronics drafters. Most of our electronics components will be housed in mechanical packages, or will be part of mechanical equipment. The number of electrical drawings in a project may be less than the mechanical drawings. In these instances, your employer will want good mechanical knowledge as well as electronics.

DIMENSIONING TERMS

You should become familiar with the following dimensioning terms:

Allowance: The amount of clearance or interference between mating parts. Example: a shaft in a wheel, with clearance, will turn freely. With interference, it will be pressed into the wheel, and they must turn together.

Tolerance: The amount a part's dimensions can vary from the basic dimension.

Unilateral tolerance: The variation allowed in only one direction. Example: $.500 \begin{smallmatrix} +.005 \\ -.000 \end{smallmatrix}$

Bilateral tolerance: The tolerance variation allowed in both directions. Example: $.500 \pm .005$

Limited dimension: The maximum and minimum sizes are specified. Example: $\begin{smallmatrix} .505 \\ .500 \end{smallmatrix}$

Fits: The tightness or looseness of mating parts. There are three general types of fits.

Clearance fits: Mating parts will always have a clearance (positive allowance).

Interference fits: Mating parts will always have interference (negative allowance).

Transition fits: A fit in which clearance or interference may occur.

DIMENSIONING RULES

The general rules for dimensioning are:

1. Each dimension must have a tolerance stated.

The tolerance can be with the dimension or stated in a general note.

2. No feature dimension should be repeated. If a dimension needs to be repeated, the second dimension must be referenced. An example is: (2.00).

3. Chain dimensioning should be avoided. It will cause a tolerance buildup. Use coordinate dimensioning. See Fig. 8-39A.

4. The finished part should be dimensioned without specifying manufacturing methods. An example would be the requirement for a 1 inch hole. If the finished hole measures 1 inch, you do not care if it was manufactured by laser beam, chemically etched, or drilled. However, engineering or manufacturing may request the process be specified.

5. Dimensions should be to visible outlines in profile views. It is not good practice to dimension to hidden lines. You should not have to look at the adjoining views to interpret a dimension.

6. Dimensions should be placed outside the view.

7. Dimensions should be placed 1/2 in. or further away from object.

8. Parallel dimensions should be spaced a minimum of 3/8 in. apart.

9. Extension lines should start 1/16 in. away from object and continue 1/8 in. beyond the dimension line.

10. Arrowheads should be three times as long as they are wide.

11. Dimension, extension, and leader lines are all .01 wide.

Fig. 8-40 demonstrates some of the rules listed above.

ANGULAR DIMENSIONS

Angular dimensions are normally shown as in Fig. 8-41. The tolerance on most angular dimensions is plus or minus one degree.

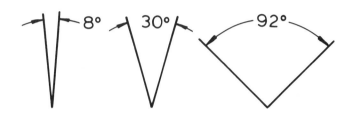

Fig. 8-41. Three examples of ways to dimension angles.

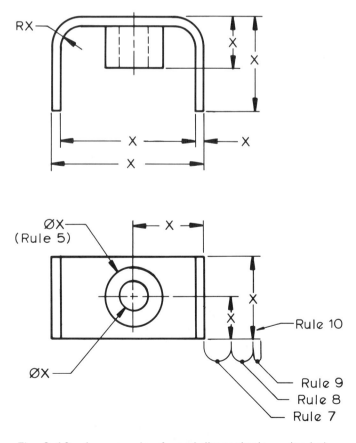

Fig. 8-40. An example of good dimensioning rules being applied.

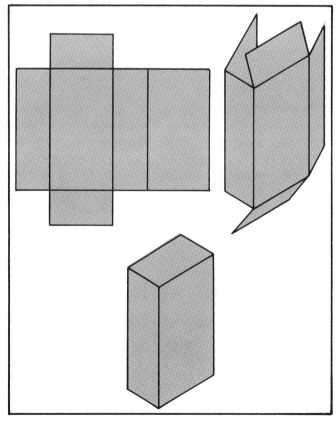

Fig. 8-42. An example of a developed stretchout being folded into a box.

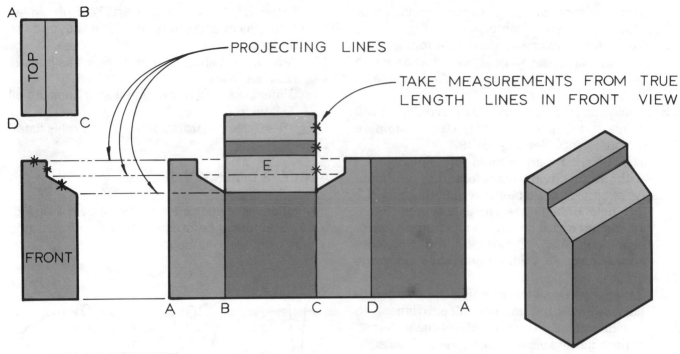

1 – DRAW TOP AND FRONT VIEWS.

2 – START THE DEVELOPMENT WITH POINT A.

3 – PROJECT THE HEIGHTS FROM THE FRONT VIEW.

4 – MEASURE TRUE DISTANCES TO INTERMEDIATE POINTS.

5 – STRETCH OUT TOP USING TRUE LENGTH MEASUREMENTS FROM THE FRONT VIEW (area E).

Fig. 8-43. Steps for developing a stretchout.

DEVELOPMENTS

Sheet metal parts are often bent into three dimensional parts. The bending is accomplished by predetermined patterns. The patterns are often called STRETCHOUTS, Fig. 8-42. The basic steps in laying out developments are shown in Fig. 8-43.

REVIEW QUESTIONS

1. Define the word packaging.
2. List four ways of permanently joining parts together.
3. What are the advantages of resistance welding?
4. List the commonly used mechanical fasteners.
5. What are the three main thread types for machine screws?
6. List the purposes for applying finishes to metals.
7. What three materials are most used in electronics packaging?
8. Compare thermosetting and thermoplastic plastics.
9. List the qualities of aluminum that make it good for electronics packaging.
10. If the inside sheet metal dimensions are critical, form the part with a _____.
    a. brake
    b. power brake
    c. die
    d. wedge
11. What does the term "bend allowance" mean?
12. What is a lightening hole?
13. In order to be a designer, a drafter must understand the _____, _____, and _____ of the part or parts.
14. How do you select a datum for dimensioning?
15. Why is a knowledge of mechanical drafting important to the electronics drafter?
16. What is rule 5 in the dimensioning practices?

PROBLEMS

PROB. 8-1. Dimension Fig. 8-44 using coordinate dimensioning. Then dimension it using the tabular system.

PROB. 8-2. Figure the bend allowance for the bend

shown in Fig. 8-45.

PROB. 8-3. Make a stretchout of Fig. 8-46. Figure the bend allowance using the following information: The material is .062 thick and is to be bent around a .12 radius.

PROB. 8-4. Add welding symbols to Fig. 8-47.
   a. A 1/4 fillet weld.
   b. A 3/16 fillet weld, all around.
   c. A 1/4 fillet weld, opposite side.
   d. A 1/4 fillet weld, all around.
   e. A 1/4 fillet weld, all around, both sides.

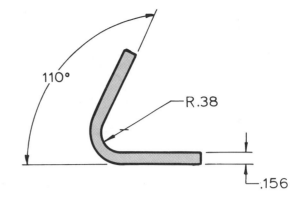

Fig. 8-45. Figure the bend allowance like that in Fig. 8-34.

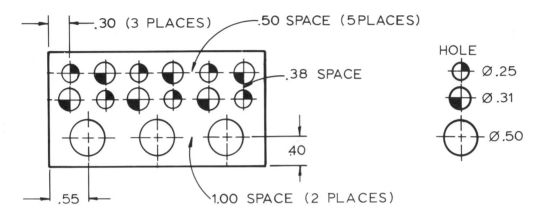

Fig. 8-44. Dimension the plate using coordinate dimensioning. Then dimension it using the tabular system.

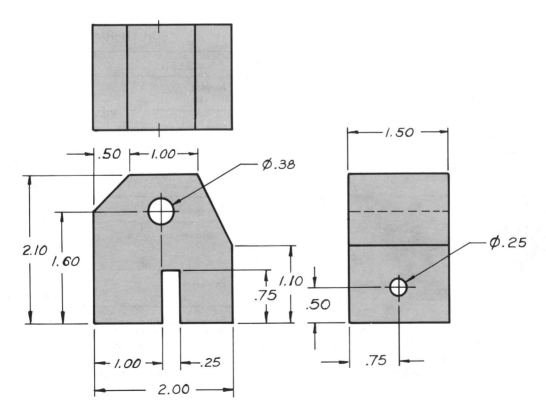

Fig. 8-46. Make a stretchout after finding the bend allowance.

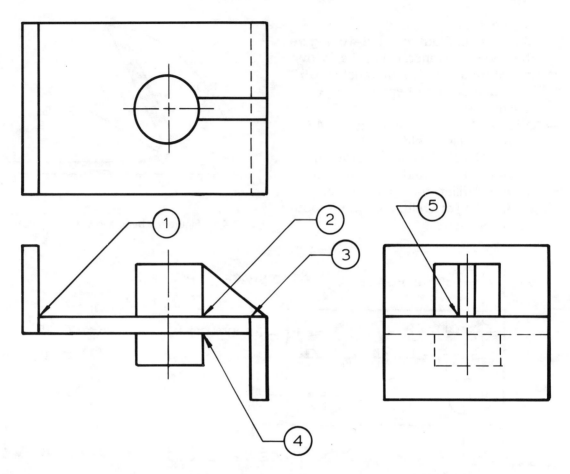

Fig. 8-47. Add fillet welding symbols as you alter the arrows.

The mechanical engineer or design drafter may design plastic enclosure prototypes. This work requires a knowledge of adhesives, resins, and molding processes. (Owens-Corning Fiberglas Corp.)

COMMON SOLVENT CEMENTS FOR THERMOPLASTICS

RESIN	SOLVENT CEMENT
Acrylic	Methylene Chloride Ethylene Dichloride
ABS	Methyl Ethyl Ketone
* Cellulosics	Methyl Alcohol Methyl Ethyl Ketone
Ethyl Cellulose	Ethylene Dichloride
Polycarbonate	Methylene Chloride
Polyphenylene Oxide	Chloroform Toluene Ethylene Dichloride
Polystyrene	Methyl Ethyl Ketone Methylene Chloride
Polyvinyl Chloride	Tetrahydrofuran

(* Cellulose acetate and cellulose acetate butyrate)

Chart of solvents to use.

Chapter 9

PICTORIAL DRAWINGS

After studying this chapter, you will be able to:
- Explain the advantages of pictorial drawings.
- Make an isometric pictorial drawing.
- Define axonometric, isometric, and oblique projection.
- Explain exploded views.
- List the steps for making isometric drawings.

Electronics drafters are often asked to create pictorial drawings. These drawings are required to direct manufacturing, marketing, maintenance, and assembly personnel. People who are untrained may not be able to read or understand multiview drawings (orthographic), but they can read pictorial drawings because they are more like the physical part. Pictorials clearly convey needed information to all readers. You will study two types of pictorial drawings: Axonometrics and Obliques.

AXONOMETRIC DRAWINGS

Axonometric drawings are pictorial drawings in which the projection lines remain parallel. This is not normally what you would see with your eyes. Your eyes would perceive the projection lines converging as they recede in the distance. See Fig. 9-1. Axonometric drawings are the most frequently used pictorials even though they are visually distorted.

There are three main types of axonometric drawings: isometric, dimetric, and trimetric, Fig. 9-2. You will study isometric drawings. These are the type electronics drafters most often use.

ISOMETRIC DRAWINGS

Isometric drawings are made with the two receding axes at 30° off the horizontal and with the third axis vertical. See Fig. 9-3. The three axis lines may be drawn and measured true length. An isometric cube will be revolved 45° and then tilted forward 35°16'. The 35°16' is the angle between

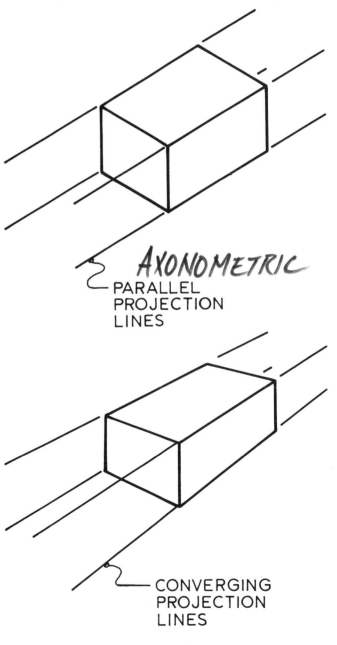

Fig. 9-1. Compare the axonometric and the natural projection. In the natural projection, the lines converge as they recede.

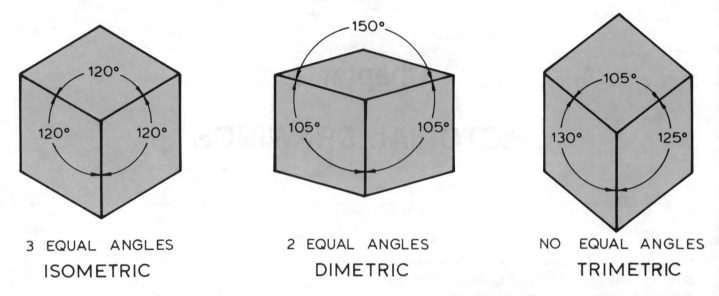

Fig. 9-2. Three main types of axonometric drawings.

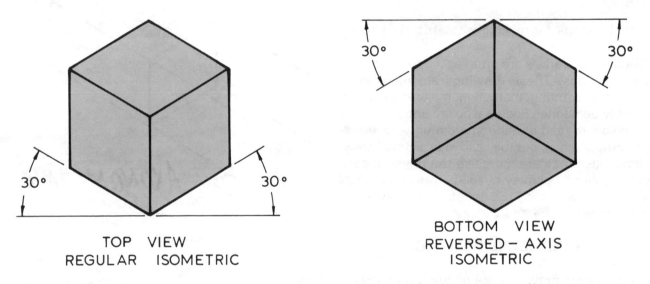

Fig. 9-3. An isometric cube drawn with regular and reversed axis.

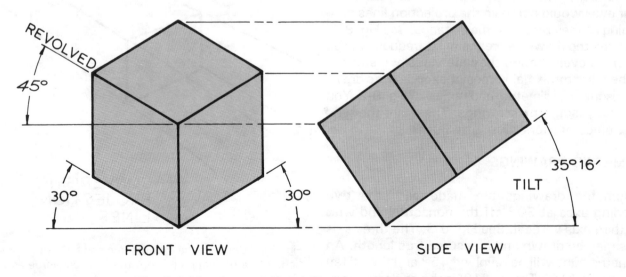

Fig. 9-4. A cube rotated and tilted for an isometric drawing.

the horizontal plane and the line of sight, Fig. 9-4.

An isometric drawing is about 1 1/4 times the size of a true representation. To show this, we will use a one inch square cube. For an isometric ellipse to fit this cube, it will have to be a 1 1/4 in. size, Fig. 9-5. If an isometric ellipse template is used, the template will have the 1 1/4 times size included in each ellipse size.

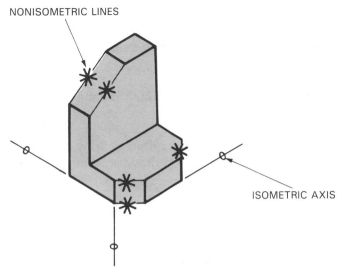

Fig. 9-7. An isometric view with nonisometric lines. These lines are not parallel to the isometric axis lines.

ISOMETRIC FROM ORTHOGRAPHIC

An isometric drawing will be based on the three orthographic views. Dimensions from the orthographic views can be transferred directly to the isometric. See Fig. 9-6. Dimensions can be transferred by using a scale or dividers.

All lines are measurable except nonisometric lines. These lines are not parallel to one of the

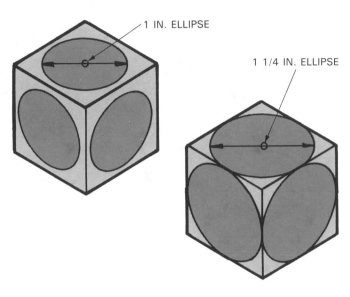

Fig. 9-5. Two isometric cubes with each axis drawn one inch long. Note that it takes a 1 1/4 in. ellipse to fit the surfaces.

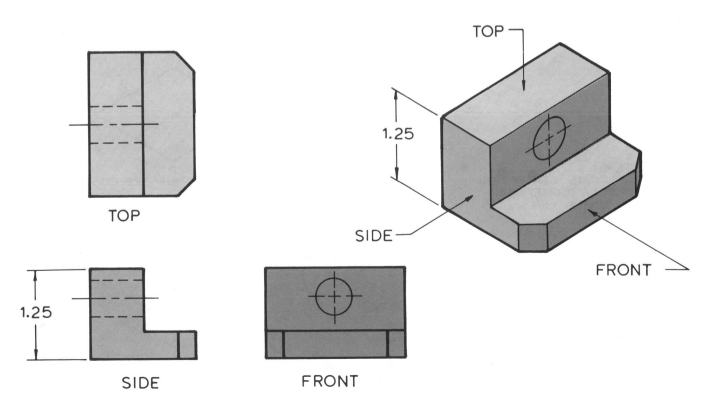

Fig. 9-6. An isometric drawing created from three orthographic views.

isometric axes, Fig. 9-7. Nonisometric lines cannot be measured, but must be plotted. Fig. 9-8 shows how to plot nonisometric lines as points from the orthographic view.

DRAWING AN ISOMETRIC ELLIPSE USING A COMPASS

The illustrator is often required to draw ellipses. Using a compass, you can draw an approximate ellipse. The compass method most often used is the FOUR-CENTER method. Fig. 9-9 shows the steps required to construct an isometric ellipse.

DRAWING AN ELLIPSE WITH TEMPLATE

An ellipse template, Fig. 9-10, is used to draw circular objects in isometric. The isometric ellipse is a 35°16' ellipse and can be used on all true isometric surfaces.

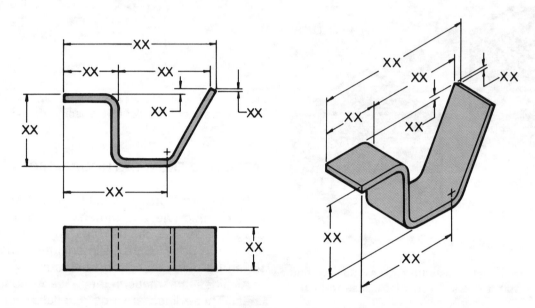

Fig. 9-8. How to plot nonisometric lines from orthographic views of oblique surfaces. Draw a box to enclose the item. Only use measurements along the directions of the edges of the box.

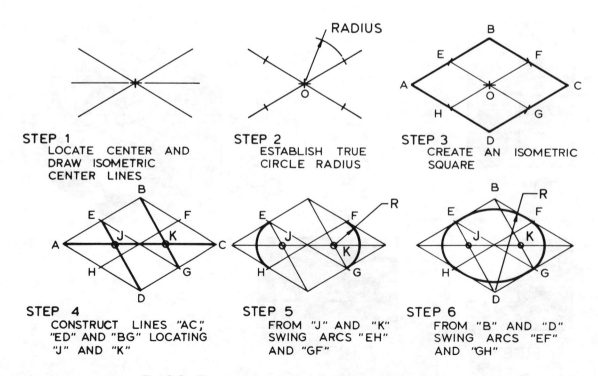

STEP 1
LOCATE CENTER AND DRAW ISOMETRIC CENTER LINES

STEP 2
ESTABLISH TRUE CIRCLE RADIUS

STEP 3
CREATE AN ISOMETRIC SQUARE

STEP 4
CONSTRUCT LINES "AC", "ED" AND "BG" LOCATING "J" AND "K"

STEP 5
FROM "J" AND "K" SWING ARCS "EH" AND "GF"

STEP 6
FROM "B" AND "D" SWING ARCS "EF" AND "GH"

Fig. 9-9. The steps for creating an ellipse using a compass.

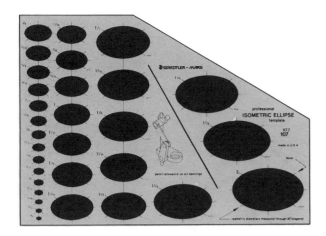

Fig. 9-10. Ellipse template.   (Staedtler—Mars)

You are also required to draw ellipses on other than true isometric surfaces. Fig. 9-11 shows how ellipse requirements change on nonisometric surfaces. An isometric protractor can be used to give us the required ellipse degree, Fig. 9-12. The isometric template may also give you the measuring value for each nonisometric line. Fig. 9-13 shows a table used on some ellipse protractors.

In order to use an ellipse template, you must learn some terms: the MAJOR and MINOR axis. The major axis is the largest measurement across the ellipse. The minor axis is the shortest distance across the ellipse. When placing the ellipse, you must be concerned about the axis. Fig. 9-14 shows how to align the ellipse on an isometric cube. The

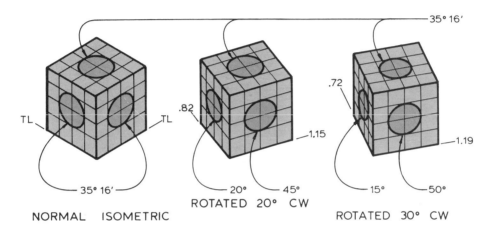

Fig. 9-11. Ellipse requirements for various viewing angles. An isometric protractor is used.

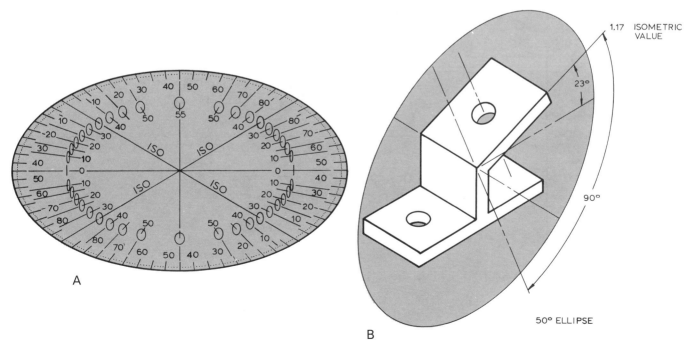

Fig. 9-12. A—An isometric protractor provides angles for drawing ellipses. B—An isometric protractor being used to establish the surface angle and ellipse requirement.

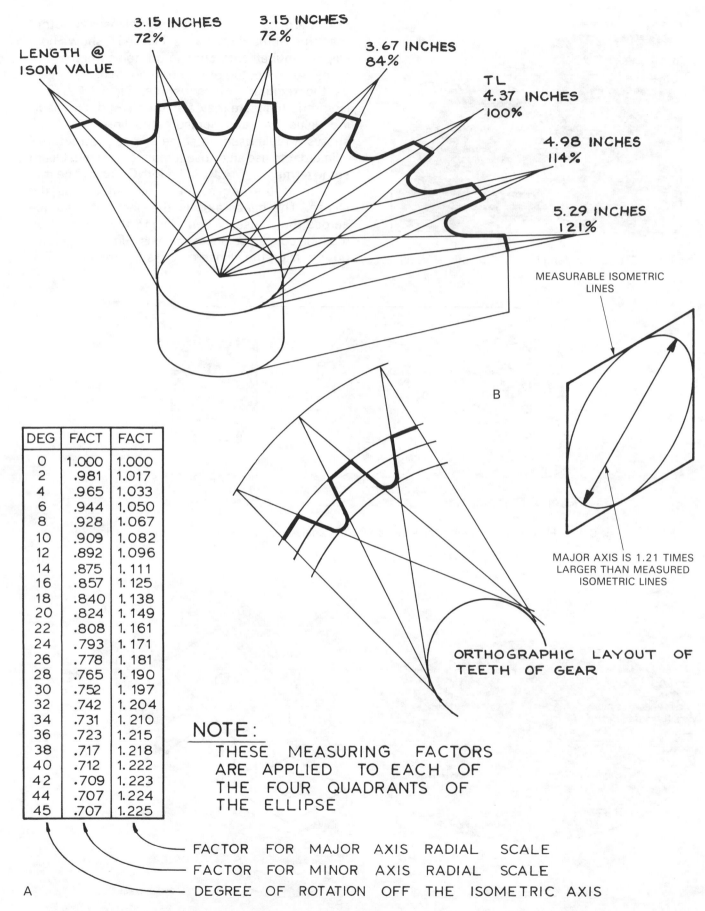

LENGTH @ ISOM VALUE

3.15 INCHES 72%

3.15 INCHES 72%

3.67 INCHES 84%

TL 4.37 INCHES 100%

4.98 INCHES 114%

5.29 INCHES 121%

MEASURABLE ISOMETRIC LINES

B

MAJOR AXIS IS 1.21 TIMES LARGER THAN MEASURED ISOMETRIC LINES

DEG	FACT	FACT
0	1.000	1.000
2	.981	1.017
4	.965	1.033
6	.944	1.050
8	.928	1.067
10	.909	1.082
12	.892	1.096
14	.875	1.111
16	.857	1.125
18	.840	1.138
20	.824	1.149
22	.808	1.161
24	.793	1.171
26	.778	1.181
28	.765	1.190
30	.752	1.197
32	.742	1.204
34	.731	1.210
36	.723	1.215
38	.717	1.218
40	.712	1.222
42	.709	1.223
44	.707	1.224
45	.707	1.225

ORTHOGRAPHIC LAYOUT OF TEETH OF GEAR

NOTE:
THESE MEASURING FACTORS ARE APPLIED TO EACH OF THE FOUR QUADRANTS OF THE ELLIPSE

—— FACTOR FOR MAJOR AXIS RADIAL SCALE
—— FACTOR FOR MINOR AXIS RADIAL SCALE
—— DEGREE OF ROTATION OFF THE ISOMETRIC AXIS

A

Fig. 9-13. A—Table shows measuring factors for each amount of rotation off the isometric axis in degrees. B—See how values apply on gear. TL means true length.

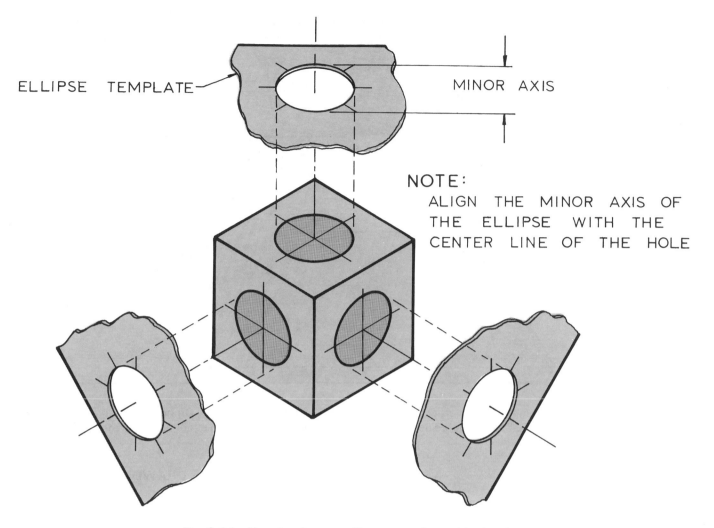

ELLIPSE TEMPLATE

MINOR AXIS

NOTE:

ALIGN THE MINOR AXIS OF THE ELLIPSE WITH THE CENTER LINE OF THE HOLE

Fig. 9-14. How to place an ellipse on an isometric drawing.

most important thing to remember is that the minor axis will align with the shaft or hole's longitudinal center line.

LONG AXIS ISOMETRIC

Long axis isometrics are drawn with the long axis horizontal and the vertical axis as a 60° angle, Fig. 9-15. When adding holes in this isometric drawing, the ellipse template will be used differently. On vertical surfaces the minor axis of the ellipse will be parallel to one 60° line. On horizontal surfaces the minor axis will be parallel to the other 60° axis.

ISOMETRIC FREEHAND SKETCHING

The initial stage for drawing isometrics is to sketch out your idea or concept. A sketch is also a valuable tool for communicating with your colleagues, engineers, and others with whom you will work. Most manufactured items start as an idea which is sketched out on paper. A simple sketch can often communicate more clearly than thousands of words. For the sketch to have full value, it must be in good proportion and include the basic detail which will appear on the finished drawing, Fig. 9-16.

STEPS FOR SKETCHING ISOMETRICS

To quickly create an isometric sketch, you should follow some organized method. Fig. 9-17 will show the steps of one adaptable method. Note how each step is required to give you total control over the final result.

DIMENSIONING ISOMETRICS

It is often required that you dimension an isometric drawing. The dimensions lend additional information and understanding for the reader. With the increased use of illustrations for manufacturing, it is necessary for you to understand dimension techniques. See Fig. 9-18.

131

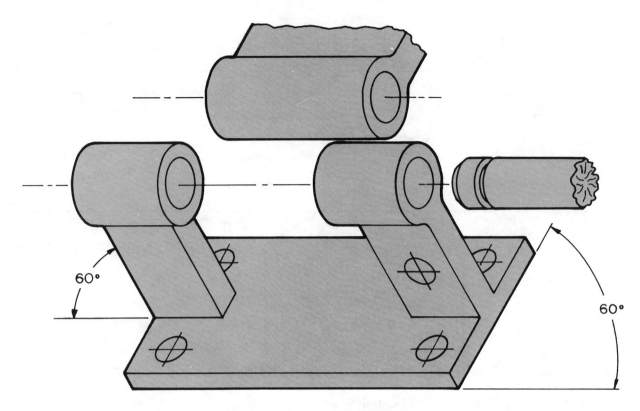

Fig. 9-15. A long axis isometric with vertical lines running at 60° off the horizontal. On horizontal surfaces, align the ellipse minor axis with 60° axis on left side of drawing. On vertical surface, align minor axis with other 60° axis.

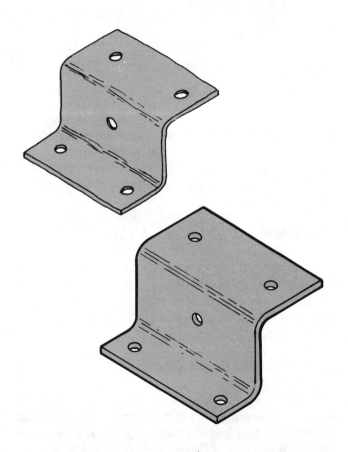

Fig. 9-16. Show the detail in a sketch when preparing for a drawing.

OBLIQUE DRAWINGS

Oblique drawings are sometimes used to show electronics equipment. The drawing is three dimensional with one surface parallel to the image plane. The other two surfaces will recede at some angle off the horizontal, Fig. 9-19. The angle is usually 30°, 45°, or 60° because of the convenience of standard triangles. All of the dimensions on the drawing's three main axes can be scaled.

SELECTING THE VIEW

The surface with the most features should be placed parallel to the image plane. This will allow you to see these features with the least amount of distortion. Fig. 9-20 shows an electronics unit drawn using this principle. The longest axis may also influence the choice of surface parallel to the image plane. The choice should minimize distortion on the long axis.

CAVALIER OBLIQUES

When an oblique drawing is laid out with the receding axis at 45°, we call it a true cavalier. General cavaliers may be drawn at any angle between 0 and 90°. See Fig. 9-21. All axis

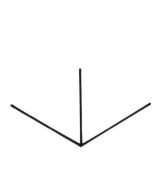

STEP 1. ESTABLISH THE ISOMETRIC AXIS

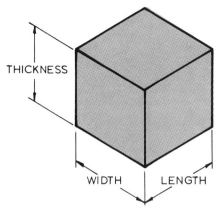

STEP 2. CREATE A CUBE WITH THE OVERALL THICKNESS, WIDTH, AND LENGTH OF THE OBJECT

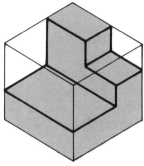

STEP 3. DRAW IN DETAILS

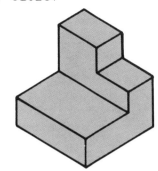

STEP 4. DARKEN IN THE DETAIL LINES AND ERASE CONSTRUCTION LINES

Fig. 9-17. There are four basic steps in creating an isometric sketch. Enclosing or ''boxing in'' an object helps avoid distorted drawings.

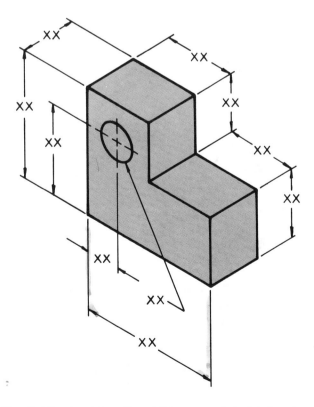

Fig. 9-18. Applying dimensions to a pictorial drawing.

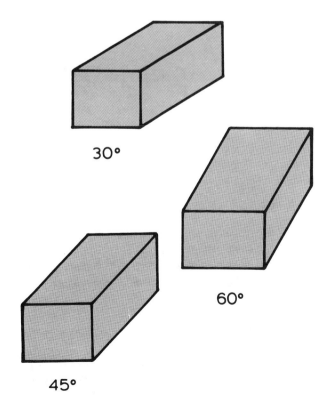

Fig. 9-19. Basic oblique drawings.

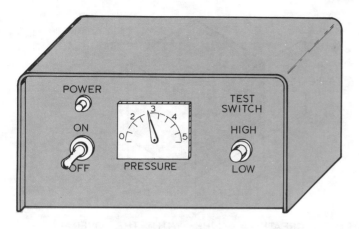

Fig. 9-20. This electronics unit is drawn in oblique so the front panel is not distorted.

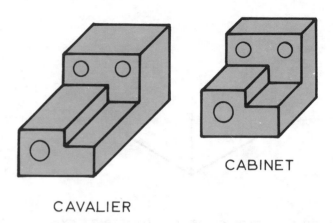

CABINET

CAVALIER

Fig. 9-22. Minimize visual distortion by drawing a cabinet oblique.

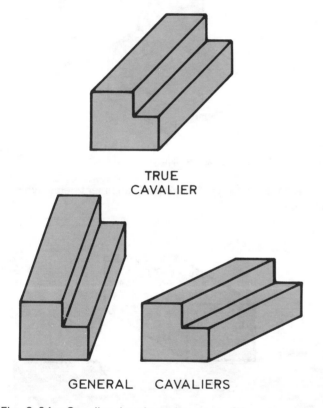

TRUE CAVALIER

GENERAL CAVALIERS

Fig. 9-21. Cavalier drawings drawn at different angles.

measurements are full measurement. This makes it easy to draw and measure the finished dimensions. A cavalier drawing distorts the object's true dimensions. To help eliminate this distortion, cabinet obliques are used.

CABINET OBLIQUES

This oblique may be drawn at any receding angle. But the receding axis will be drawn 1/2 size. See Fig. 9-22. This shortening of the receding axis will minimize the distortion seen in the cavalier.

HOLES IN OBLIQUE DRAWINGS

It is possible to use templates or a compass to create holes or cylindrical shapes on oblique drawings. The front surface is easiest, as all the shapes are true. But circles on receding surfaces will need to be designed. You can use the four-center method, Fig. 9-23. Or you can establish the hole's "Y" and "Z" dimensions so an ellipse template may be used, Fig. 9-24.

EXPLODED DRAWINGS

Exploded drawings are used extensively for assembly and maintenance documents. They show

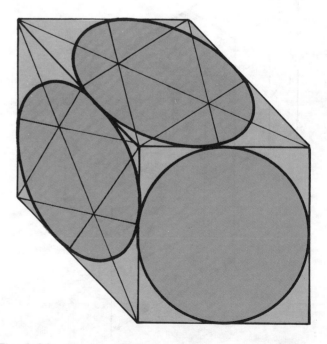

Fig. 9-23. Drawing a hole in an oblique surface using the four-center method.

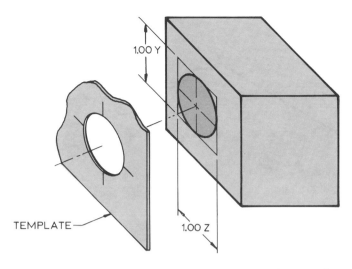

Fig. 9-24. To draw hole on oblique drawing, lay out the Y and Z dimensions. Then select an ellipse that fits best.

how the individual parts fit together, Fig. 9-25. Each part will be numbered so it can be found in a parts list or parts catalog. The parts list or parts catalog will give a full description for each part.

Often small features or parts will have to be shown more clearly than the scaled drawing allows. A detail of the part or feature will be needed. Fig. 9-26 shows a method used by many illustrators.

REVIEW QUESTIONS

1. What are some uses for pictorial drawings?
2. Why is freehand sketching important?
3. List three axonometric drawings.
4. What is meant by boxing in?
5. List steps for sketching an isometric.
6. Define isometric drawing.

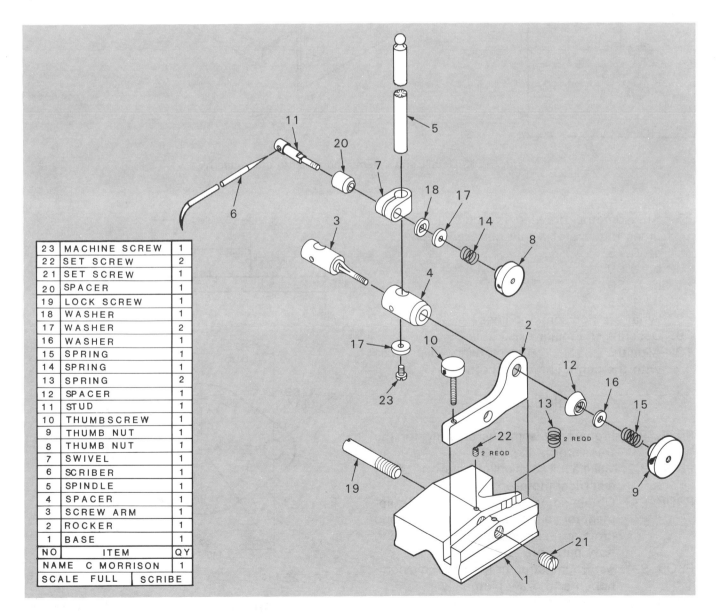

NO	ITEM	QY
23	MACHINE SCREW	1
22	SET SCREW	2
21	SET SCREW	1
20	SPACER	1
19	LOCK SCREW	1
18	WASHER	1
17	WASHER	2
16	WASHER	1
15	SPRING	1
14	SPRING	1
13	SPRING	2
12	SPACER	1
11	STUD	1
10	THUMBSCREW	1
9	THUMB NUT	1
8	THUMB NUT	1
7	SWIVEL	1
6	SCRIBER	1
5	SPINDLE	1
4	SPACER	1
3	SCREW ARM	1
2	ROCKER	1
1	BASE	1
NO	ITEM	QY
NAME	C MORRISON	1
SCALE FULL	SCRIBE	

Fig. 9-25. An exploded drawing used for assembly.

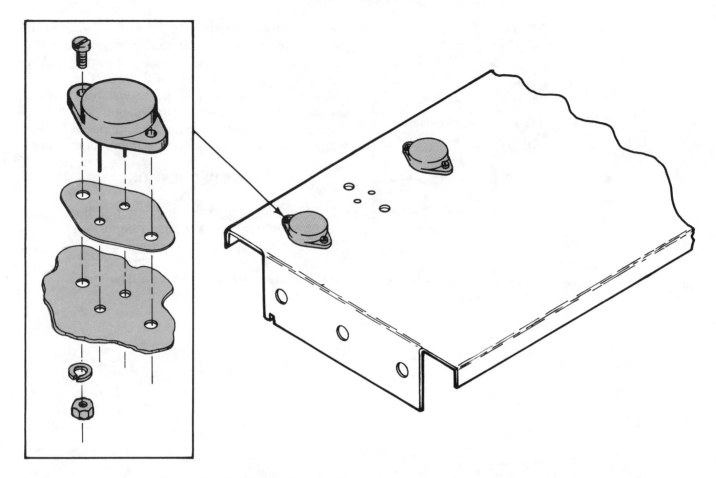

Fig. 9-26. Details show items too small to draw clearly.

7. An isometric drawing is about _____ times the size of a true representation.
   a. 0.75
   b. 0.90
   c. 1.50
   d. 1.25
8. What are nonisometric lines?
9. Describe the major axis.
10. Align the _____ (major, minor) ellipse axis with the centerline of a shaft.

PROBLEMS

PROB. 9-1. Prepare an isometric drawing for the transistor in Fig. 9-27.

PROB. 9-2. Make a full scale oblique drawing of the electrical motor in Fig. 9-28.

PROB. 9-3. Obtain some broken electronic equipment for your class. Take this apart and draft an exploded isometric showing how parts are assembled.

PROB. 9-4. Ask your instructor for electronic components to draw pictorially.

PROB. 9-5. Create a cabinet oblique drawing of an electronic device.

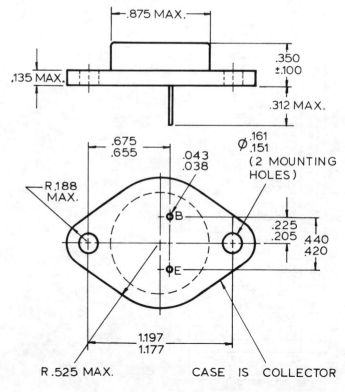

Fig. 9-27. A typical power transistor body. Prepare an isometric drawing.

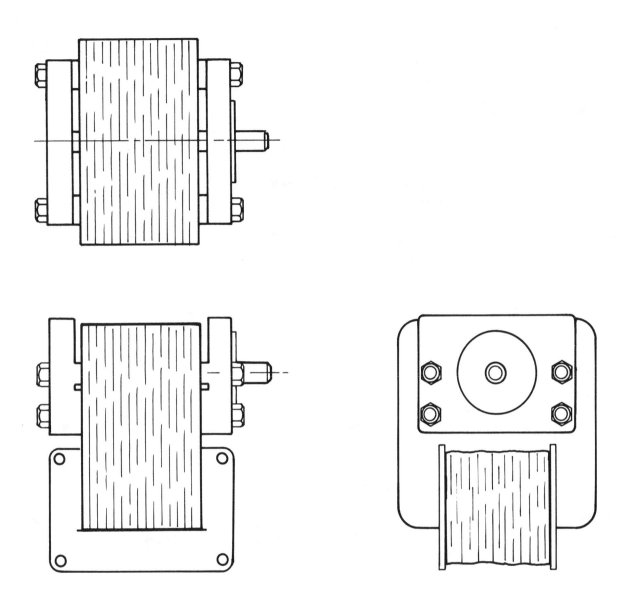

Fig. 9-28. An alternating current motor. Make an oblique drawing.

A computer display terminal and an interactive digital plotter being used to design an electronic circuit. (Tektronix, Inc.)

Chapter 10

ELECTRONIC DRAFTING USING AUTOCAD*

At the completion of this chapter you will be able to:
- Create electronic symbols.
- Save symbols for future use.
- Insert symbols into a drawing.
- Use the STRETCH command to arrange a schematic.
- Draw schematics using AutoCAD software.

Most electronics firms use some kind of a computer-aided drafting system to create their drawings. This chapter discusses the basic method for using AutoCAD for drawing in electronics. *It is assumed that you already know how to use AutoCAD, and that you are able to generate simple mechanical drawings.*

SETUP

Industrial practices will dictate your computer setup. Units of measurement for your drawings have been established by the International Standards Organization (ISO) and the American National Standards Institute (ANSI). Electronic components are manufactured all around the world. When you design a circuit board for a memory chip, you do not know if the chip is going to be manufactured in Malaysia, Hong Kong, Mexico, or the United States. The components from all these countries are interchangeable.

Turn on your computer and establish the following setting:

Command: **GRID** ↵
Grid spacing (X) or ON/OFF/Snap/Aspect ⟨0⟩: **1** ↵

Note: If no units have been set in the computer previously, inches are used as default units.

Setting the grid to 1 in. allows you an opportunity to keep its relative size in mind.

Command: **SNAP** ↵
Snap spacing or ON/OFF/Aspect/Rotate/Style ⟨current⟩: **.1** ↵

This snap setting helps you *pick* features more easily on your electronics drawings.

Command: **UNITS** ↵
Number of digits to the right of decimal point (0 to 8)⟨4⟩: ↵
System of angle measure:
    Enter choice, 1 to 5,⟨1⟩: ↵
    Number of fractional places for display of angle (0 to 8)⟨0⟩: ↵
Direction for angle 0:
    Enter direction angle 0⟨0⟩: ↵
    Do you want angles measured clockwise? ⟨**N**⟩ ↵

Press F1 to get back to the drawing editor.

AutoCAD RELEASE 11

Text will appear on your screen explaining AutoCAD's unit options. For your electronic drafting purposes, choose decimal units.

    Enter choice, 1 to 5: **2** ↵ **(decimal units)**
    Command: **BLIPMODE** ↵
    ON/OFF ⟨current⟩: **OFF** ↵
    Command: **REGENAUTO** ↵
    ON/OFF ⟨current⟩: **OFF** ↵

Setting the BLIPMODE and the REGENAUTO off will help avoid clutter on your drawing. Also, the drawing will regenerate less frequently, saving you time.

    Command: **ORTHO** ↵
    ON/OFF: **ON** ↵

Turning ORTHO on allows you to easily create straight horizontal or vertical lines. This is definitely a time-saver. In addition, most lines on your electronics drawings run on the horizontal or vertical axes.

This completes the setup. Now you are ready to create some symbols.

CREATING AN ELECTRONIC SYMBOLS LIBRARY

When creating a symbol, set it up so that it enters into your drawing ready to connect onto the existing lines. Also, the lines should attach to the ends of the symbols easily. This is the reason that snap is set to .1 in. A .1 in. snap is about the smallest snap that will work effectively. Move your cursor on the screen. Note how the cursor snaps 10 times between grid points. Draw a resistor symbol using the following steps:

Note: Be sure that the status line indicates that your snap is on. If not, push the F9 function button and your snap will toggle on.

    Command: **LINE** ↵
    From point: **(pick point 1 — the start of the line will be an attachment point for the symbol)**
Next, type the following for points 2 through 8:
    To point: **@.0833⟨60** ↵
    To point: **@.1666⟨300** ↵
    To point: **@.1666⟨60** ↵
    To point: **@.1666⟨300** ↵
    To point: **@.1666⟨60** ↵
    To point: **@.1666⟨300** ↵
    To point: **@.0833⟨60** ↵
    To point: ↵

Both ends of your symbol will start and stop on a .1 in. snap. All symbols that you draw should have their attachment points set up this way. It makes your symbols easy to connect with other components.

Notice you are leaving the tails off the resistor symbol. The tails were put on the symbol when you used your templates during board drafting, because they helped the drafter align the symbol with existing lines on the drawing. Here they are best left off, Fig. 10-1. Your plotted drawings look better when you change colors to differentiate your elements.

Fig. 10-1. Resistor symbols with tails, as are drawn with a template, and resistor symbols without tails, as are drawn with AutoCAD.

CREATING A BLOCK FOR SYMBOLS

After you create a symbol, you may want to use it several times in your drawing. To do this, you must make the symbol into a block. Follow these steps to make your resistor symbol a block:

Command: **BLOCK** ↵
Block name (or ?): **R** ↵ **(keep your block names simple)**
Insertion base point: **'ZOOM** ↵
⟩⟩Center/Dynamic/Left/Previous/Window/⟨Scale(X)⟩: **W** ↵
⟩⟩First corner: **(pick on lower-left side of symbol)** ⟩⟩Other corner: **(pick on upper-right side of symbol)**
Insertion base point: **(pick point 1)**
Select objects: **W** ↵
First corner: **(pick on lower-left side of symbol)** Other corner: **(pick on upper-right side of symbol)**
Select objects: ↵

Note: All of the parts of your symbol will disappear. This is supposed to happen.

You now have a symbol that may be placed into this drawing. However, if you wish to use it in other drawings, you must write it to a separate file. This is done by creating a wblock.

CREATING A WBLOCK FOR SYMBOLS

A wblock will allow you to use your symbol in any future drawing. Creating a wblock writes your symbol information to a permanent file. To put your resistor on file so that it can be used in other drawings, follow this procedure:

Command: **WBLOCK** ↵
File name: **R** ↵
Block name: **=** ↵

This gives the wblock the same name that was used for your block.

It is not necessary for the names to be the same, but it is advisable. Also, your hard disk is chosen as the default drive to store your wblock. If you desire to store your wblock on a floppy disk, you would alter the previous sequence by:

File name: **A:R** ↵ **(A can be substituted with any drive letter of your choice)**

Once the symbol has been written to the AutoCAD file you may use it in any drawing. You can use the same basic procedure to create other symbols. Most of the symbols you will need for this book's exercises are shown in Fig. 10-2.

USING SYMBOLS TO CREATE ELECTRONICS DRAWINGS

Now that you have learned to build electronics symbols, the next step is to create an electronics drawing with the symbols. To create your electronics drawing you must be able to place and alter your symbols.

INSERTING YOUR SYMBOLS

To insert symbols, or blocks as they are referred to in AutoCAD, into your drawing, follow these steps:

Command: **INSERT** ↵
Block name (or ?): **R** ↵ **(for your resistor symbol)**
Insertion point: **(pick where you want your resistor placed)**
X scale factor ⟨1⟩/Corner/XYZ: ↵
Y scale factor ⟨default = X⟩: ↵

Normally, you will have drawn the symbols the size you want them.

Rotation angle ⟨0⟩: ↵

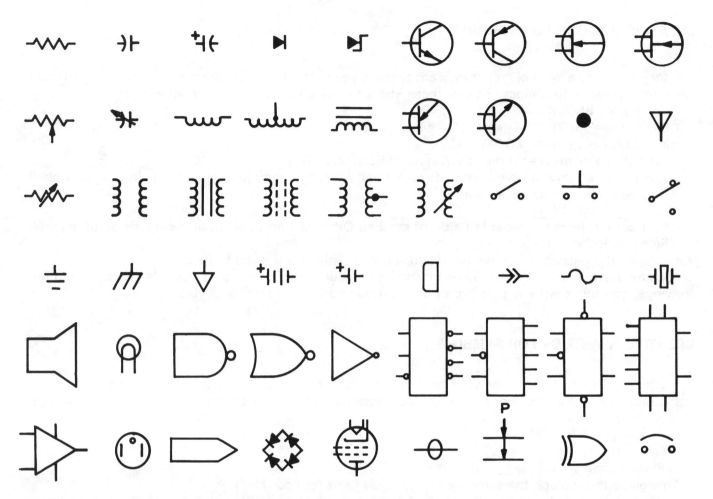

Fig. 10-2. These are many of the common schematic symbols. Try to produce some of these with AutoCAD.

Generally, in electronics drafting, you will only rotate electronic symbols at some multiple of 90 degrees. If you need a symbol rotated, type the desired degree and press return.

The computer will now enter the symbol into your drawing. Fig. 10-3 demonstrates INSERT options.

If you are using a lot of one kind of component on your schematic and the components are arranged in a designated pattern, you may chose to copy them with the ARRAY command. This way you do not have to work through the INSERT command so often. If they are not systematically arranged, use the COPY command. Once the components have been copied, you can use the normal editing commands to place them where they are needed.

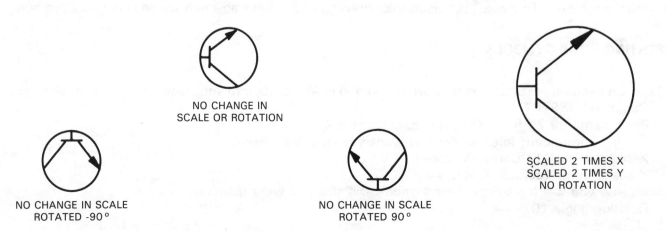

Fig. 10-3. Using the insert options, your symbols can be adapted to many situations without redrawing them.

ALTERING SYMBOLS

Many schematic symbols in electronics are very similar. Thus, it is often faster to alter a symbol you have already made, than to start from scratch. To explore this procedure, first you will build your resistor into a potentiometer (variable resistor), and then you will alter the symbol to create a tapped resistor. Begin with your resistor on screen:

Command: **PLINE** ↵

This command allows you to build lines of varying width.

From point: **(pick the tip of one of the upper peaks in your resistor)**
Current line-width 0.0
Arc/Close/Halfwidth/Length/Undo/Width/⟨Endpoint of line⟩: **W** ↵
Starting width ⟨0.000⟩: ↵
Ending width ⟨0⟩: **.125** ↵
Arc/Close/Halfwidth/Length/Undo/Width/⟨Endpoint of line⟩: **(pick a point two snaps above your starting point)**

You have just built the point of the arrow for the potentiometer. To finish the potentiometer, pick a short vertical line on top of the arrowhead. The resulting figure should look similar to the potentiometer in Fig. 10-4. Now make your potentiometer a block and a wblock using the commands previously outlined.

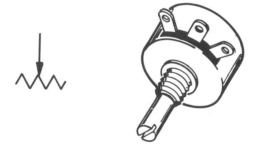

Fig. 10-4. The potentiometer schematic you have just drawn should look similar to the one above. Along side is a picture of an actual potentiometer.

To change your new symbol into a tapped resistor, begin with the potentiometer block on screen and do the following:

Command: **EXPLODE** ↵
Select block reference, polyline, or dimension: **(pick the potentiometer)**
Command: **ERASE** ↵
Select objects: **(pick the arrowhead)** ↵
Select objects: **(pick the small vertical line) (return)**
Select objects: ↵

The EXPLODE command allows your block to be taken apart piece-by-piece. You can then pick the information that needs to be altered. Now, you can make your alterations. First, you want to put a small circle where the arrowhead had been.

Command: **DONUT** ↵
Inside diameter ⟨current⟩: **0** ↵
Outside diameter ⟨current⟩: **.05** ↵
Center of doughnut: **(pick a point so that the small dot will rest on one of the resistor's peaks)**

To finish the tapped resistor, draw a short vertical line on top of the small dot you have just created. This completes the transformation to the tapped resistor, Fig. 10-5. Save this symbol as a block and a wblock.

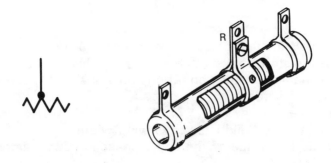

Fig. 10-5. Your tapped resistor schematic should look like the one above. Along side is a picture of an actual tapped resistor.

THE STRETCH COMMAND

The STRETCH command is very useful in electronic drawings. STRETCH makes it easy to space your components nicely on the paper. Fig. 10-6 shows an unedited, crowded schematic. Fig. 10-7 shows the same schematic after stretching and adding components. The following demonstrates use of the STRETCH command.

Command: **STRETCH** ↵
Select objects to stretch by window . . .
Select objects: **C** ↵ **(the C chooses the crossing box)**
First corner: **(pick the first corner)**
Other corner: **(pick the second corner)**
Select objects: **(press return unless you wish to pick any additional objects)**
Base point: **(pick the beginning point of the stretch)**
New point: **(pick the ending point of the stretch)**

This command saves time for drafters when making changes. Frequently, engineers will want to add or delete components. This will require rearranging large portions of the schematic. In such instances the STRETCH command proves very helpful.

PLOTTING THE DRAWING

It is important to produce a good-looking and easy-to-read copy of your drawing. People working with your drawings must be able to follow what is, often, a very complex schematic. Plotting different elements of the schematic with different colors, or layers, is an effective method to help make your printout understandable. Then, an appropriate and complete plot must be produced.

CREATING NEW LAYERS

The LAYERS command is used frequently by the electronics drafter. It allows you to set up the plotter and the monitor for color and linetype emphasis. To create a new layer:

Command: **LAYER** ↵
?/Make/Set/New/ON/OFF/Color/Ltype/Freeze/Thaw: **N** ↵
New layers name(s): **SYMBOLS** ↵
?/Make/Set/New/ON/OFF/Color/Ltype/Freeze/Thaw: **C** ↵
Color: **1** ↵

This gives the current layer the title, SYMBOLS. All elements of this layer are drawn in color 1 (red). Using the preceding procedure, create a layer in white for the connecting lines, called LINES, and one for any textual information, called TEXT, in green. This will help you visually organize your schematic.

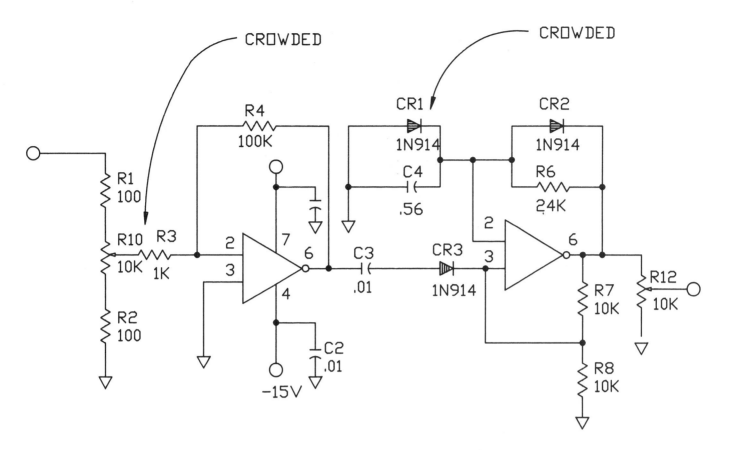

Fig. 10-6. Notice how some components are crowded together. Also, there is no room to add additional information or components.

SETTING LAYERS

Setting a layer makes it the current layer, the layer you will be working on. If you have the SYMBOLS layer on you will be drawing on a layer that is red in color. Here are the steps for setting your layers:

    Command: **LAYER** ↵

    ?/Make/Set/New/ON/OFF/Color/Ltype/Freeze/Thaw: **S** ↵

    New current layer ⟨current⟩: **LINES** ↵

This changes the current layer to LINES. Be careful to have the desired layer on when you insert the symbols — they are inserted into the drawing on the current layer. So, if you want red symbols, you should have the SYMBOLS layer on while you insert them. If the LINES layer is on, the symbols will come into the drawing in white.

Another way to bring symbols correctly into any layer is to set the color command to BYBLOCK. The symbols will then come into the drawing with their drawn entities. In other words, if they are drawn red you may insert them into a white layer and they will still be red.

TURNING LAYERS ON AND OFF

If a layer is turned off, it will not show on the screen. This can be an advantage when you are trying to eliminate some of the confusion on cluttered drawings. It also will speed up your computer's operation, since it will not have to deal with the layer that is turned off. When your drawing is plotted, only layers that are turned on will be shown. To turn layers on or off:

    Command: **LAYER** ↵

    ?/Make/Set/New/ON/OFF/Color/Ltype/Freeze/Thaw: **ON (or) OFF** ↵

When designing printed circuit boards it is common to have different layers for the board, circuit-side circuitry, component-side circuitry, component designations, dimensioning, notes, and any other items

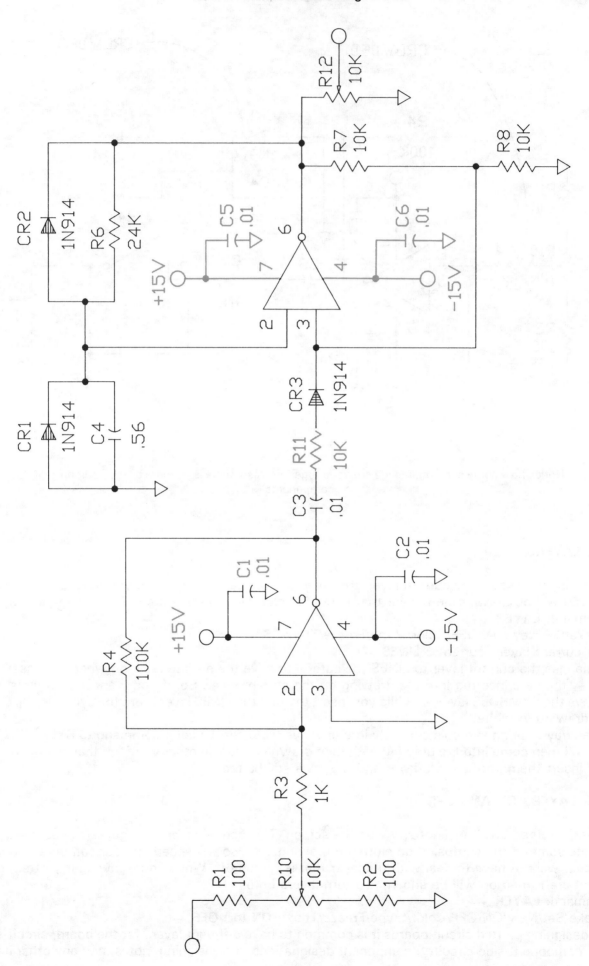

Fig. 10-7. The same schematic as Fig. 10-6 after the STRETCH command has been used. The components are spaced more evenly and additional information and components have been inserted.

required by the drawing. You can turn layers on and off to make more than one plotted drawing from the same computer drawing. For example, you could have the board and the dimension layers turned on to create a drawing to manufacture the board. The plotted drawing will give the board's manufacturer just the information needed for production.

PLOTTING DRAWINGS

You should use the ZOOM command with the Extents (E) option before you go to the plotter with your drawing. This will assure you that you do not have something drawn outside your screen viewing area. If the drawing looks good, you should save it. Always save your drawing before going to the plotter or printer. This way, if there is a problem in transmission, you will not lose your work. It is a shame to lose your drawing after all the work is accomplished.

Now you are ready to plot. Here is an example of a PLOT setup for an electronics drawing:

Command: **PLOT** ↵

What to plot-Display, Extents, Limits, View or Window ⟨D⟩: **D** ↵

This option plots the current display. The following text is then printed.

Plot will NOT be written to a selected file

Sizes are in Inches

Plot origin is at (0.00,0.00)

Plotting area is XXX wide by XXX high (MAX size)

Plot is NOT rotated 90 degrees

Pen width is 0.010

Area fill will be adjusted for pen width

Hidden lines will NOT be removed

Plot will be scaled to fit available area

Do you want to change anything? ⟨N⟩

Pressing the return key will initiate the plot using the stated defaults. If you wish to change any of the defaults hit Y and then return. Fig. 10-8 will appear on screen. If you wish to alter pen numbers, linetypes, or pen speeds you may do so here.

```
Enter values   blank=Next value, Cn=Color n, S=Show current values
X=Exit

Entity     Pen     Line     Pen        Entity     Pen     Line     Pen
Color      No.     Type     Speed      Color      No.     Type     Speed

1 (red)     1       0        36         9          1       0        36
2 (yellow)  2       0        36        10          2       0        36
3 (green)   3       0        36        11          3       0        36
4 (cyan)    4       0        36        12          4       0        36
5 (blue)    5       0        36        13          5       0        36
8 (magenta) 8       0        36        14          6       0        36
7 (white)   7       0        36        15          7       0        36
8           8       0        36
Line types:         0 = continuous line

                    1 = ...............................

                    2 = _____

                    3 = _ _ _ _ _ _ _ _ _ _ _ _ _ _ _

                    4 = _ . _ . _ . _ . _ . _ . _ . _

                    5 = __ __ __ __ __ __ __ __ __ __

                    6 = __ _ __ _ __ _ __ _ __ _ __ _

Do you want to change any of the above parameters? (N)
```

Fig. 10-8. This screen shows your current pen number, linetype, and pen speed settings. All of them can be adjusted to your design needs.

The rest of your options are next presented for you to alter. Some common changes include:

Size units (Inches or Millimeters) ⟨current⟩: **(Make sure you are using the correct unit system.)**

Adjust area fill boundaries for pen width? ⟨N⟩: **(When working with printed circuit boards, where extreme accuracy is needed, you may want to hit Y. This offsets the pens by one half the pen width, for a more accurate plot.)**

Specify scale by entering:

Plotted units = Drawing units or Fit or ? ⟨default⟩: **(Typing an F here will fit your drawing to the size paper you are using. If you have specific size requirements, you may type them here. If you need a scale of 1 in. on your plot to 3 in. on your drawing, you would type 1 = 3 at the prompt.)**

REVIEW QUESTIONS

1. What snap do you use when creating electronic drawings?
2. Why do you use this snap setting?
3. List the steps in creating a block.
4. How does a wblock differ from a block?
5. How do you change the size of a symbol while it is being inserted?
6. What does the EXPLODE command do?
7. The STRETCH command enables you to rearrange your schematic. List the steps required to stretch an area on your drawing.
8. What are the advantages of turning off layers of your drawing?
9. What should be done before plotting your drawing?

ACTIVITIES

1. Create any symbols or components assigned by your instructor. They all should have a snap of .1 in.
2. Save your new symbols and components as blocks.
3. Make wblocks out of the blocks you have created.
4. INSERT your blocks into a drawing, and test them for accuracy.
5. Plot the work assigned by your instructor.

REFERENCE INFORMATION

RECOMMENDED MINIMUM BEND RADII FOR ALUMINUM ALLOYS

Thickness	3003-O 5052-O 5052-H32 6061-O	3003-14 C 2014-O 2024-O C 2219-O 5052-H32 6061-T4 C 7075-O	2014-O 7075-O	6061-T6	C 2014-T6 C 2219-T81 C 7075-T6	2024-T3 C 2024-T3 C 2014-T3	2024-T36 C 2024-T36
.012	.03	.03	.03	.03	.09	.09	.09
.016	.03	.03	.03	.03	.09	.09	.09
.020	.03	.03	.06	.03	.12	.09	.09
.025	.03	.03	.06	.03	.16	.12	.12
.032	.03	.03	.06	.06	.19	.12	.12
.040	.06	.06	.09	.09	.25	.16	.16
.050	.06	.06	.12	.09	.31	.19	.19
.063	.06	.09	.12	.12	.41	.19	.25
.071	.06	.12	.19	.16	.44	.22	.28
.080	.09	.12	.22	.16	.50	.31	.44
.090	.09	.16	.25	.19	.56	.38	.47
.100	.09	.19	.25	.22	.66	.41	.53
.125	.12	.22	.38	.28	.84	.50	.62
.160	.16	.28	.53	.34	1.06	.66	.75
.188	.19	.38	.66	.47	1.38	.84	.91
.250	.28	.62	1.00	.75	2.00	1.25	1.25
.375	.41	1.38	1.50	1.50		1.88	1.88
.500	.69	2.50	2.50	2.50		2.50	2.50

*"C" on Material types indicates clad material

RECOMMENDED MINIMUM BEND RADII FOR STEEL ALLOYS

THICKNESS	SAE ALLOYS	
	950 1010 1020 1025	4130
.020	.06	.06
.025	.06	.06
.032	.06	.09
.036	.06	.09
.040	.06	.12
.050	.09	.12
.063	.12	.16
.080	.16	.22
.090	.19	.25
.112	.22	.28
.125	.25	.31
.160	.31	.41
.188	.38	.47
.250	.50	.62
.375	.75	.94

MATHEMATICAL VALUES

Prefix	Multiplier	Symbol
Kilo	10^3	K*
Milli	10^{-3}	m
Mega	10^6	M
Micro	10^{-6}	U or Y
Giga	10^9	G
Nano	10^{-9}	N
Tera	10^{12}	T
Pico	10^{-12}	P or UU

*Most often hand-lettered, capital; i.e. **K**

STANDARD ABBREVIATIONS

ACTUATOR	**ACTR**	CYCLES PER MINUTE	**CPM**	OHMMETER	**OHM**
ADAPTER	**ADPTR**	CYCLE PER SECOND	**CPS**	ORANGE	**ORN**
ADJUST	**ADJ**	DIAMETER	**DIA**	OSCILLATOR	**OSC**
ALARM	**ALM**	DIMENSION	**DIM.**	PANEL	**PNL**
ALIGNMENT	**ALIGN.**	DIODE	**DIO**	PART	**PT**
ALLOWANCE	**ALLOW.**	DIRECT CURRENT	**DC**	PART NUMBER	**PN**
ALTERNATING CURRENT	**AC**	ELECTRIC	**ELEC**	POSITIVE	**POS**
ALTERNATOR	**ALTNTR**	EQUIPMENT	**EQPT**	POTENTIOMETER	**POT**
ALUMINUM	**AL**	FARAD	**F**	POWER	**PWR**
AMMETER	**AMM**	FREQUENCY	**FREQ**	PURPLE	**PRP**
AMPERE	**AMP**	GAGE	**GA**	RECEIVER	**RCVR**
AMPLIFIER	**AMPL**	GENERATOR	**GEN**	RECEPTACLE	**RCPT**
ANTENNA	**ANT**	GRAY	**GRA**	RECTIFIER	**RECT**
BATTERY	**BAT**	GROMMET	**GROM**	RESISTOR	**RES**
BLUE	**BLU**	GROUND	**GND**	SCHEMATIC	**SCHEM**
BOARD	**Bd**	HARDWARE	**HDW**	SINGLE POLE	**SP**
BOTTOM	**Bot**	INDICATOR	**IND**	SINGLE POLE DOUBLE THROW	**SPDT**
BRACKET	**BRKT**	INDUCTANCE COIL	**L**	SINGLE POLE SINGLE THROW	**SPST**
BROWN	**BRN**	INSTALLATION	**INSTL**	SOLENOID	**SOL**
CADMIUM	**CAD.**	KILO	**K**	SYMBOL	**SYM**
CAPACITOR	**CAP.**	KILOHM	**K**	TERMINAL	**TERM**
CATHODE	**C**	KILOVOLT	**KV**	TRANSFORMER	**XFMR**
CATHODE-RAY TUBE	**CRT**	KILOWATT	**KW**	TRANSISTOR	**XSTR**
CENTER	**CTR**	LIGHT	**LT**	TRANSMITTER	**XMTR**
CHASSIS	**CHAS**	MAINTENANCE	**MAINT**	VIOLET	**VIO**
CHECK	**CHK**	MANUFACTURE	**MFR**	VOLT	**V**
CIRCUIT	**CKT**	MATERIAL	**MATL**	VOLTAGE REGULATOR	**VR**
CIRCUIT BREAKER	**CB**	MAXIMUM	**MAX**	VOLTMETER	**VM**
COAXIAL	**COAX.**	MEASURE	**MEAS**	WATT	**W**
COLOR CODE	**CC**	MEGA (10^6)	**MEG**	WHITE	**WHT**
CONDUCTOR	**CNDCT**	METER	**MTR**	YELLOW	**YEL**
CONNECTOR	**CONN**	MICRO (10^{-6})	**U**	WIRE	**W**
CONSOLE	**CSL**	MICROFARAD	**UF**	WIRING, TIEPOINT	**WR**
COPPER	**COP**	MICROHENRY	**UH**	SOCKET	**X**
CRYSTAL	**XTAL**	MILLI (10^{-3})	**M**	FUSEHOLDER (socket)	**XF**
CURRENT	**CUR.**	MILLIAMPERE	**MA**	CRYSTAL, PIEZOELECTRIC	**Y**
		NEGATIVE	**NEG**	OSCILLATOR	**Y**

STANDARD REFERENCE DESIGNATIONS

AR	Amplifier	**H**	Hardware	**Q**	Rectifier (transistor)
B	Blower	**HR**	Heater		Transistor
	Fan	**HT**	Headset	**R**	Potentiometer
BT	Battery	**J**	Connector (receptacle)		Resistor
C	Capacitor		Jack		Rheostat
CB	Circuit breaker	**K**	Relay	**RT**	Resistor (thermal)
CR	Diode (crystal)		Relay (solenoid)	**S**	Dial, telephone
	Rectifier (diode)	**L**	Choke coil		Key (telegraph)
	Rectifier (metallic)		Coil		Switch
DS	Alarm		Inductor	**T**	Autotransformer
	Annunciator		Solenoid (electrical)		Transformer
	Buzzer		Winding	**TB**	Block (connecting)
	Indicator	**LS**	Horn		Terminal board
	Lamp (fluorescent)		Loudspeaker		Test block
	Lamp (incandescent)		Speaker	**TP**	Test point
	Neon Lamp	**M**	Clock	**U**	Integrated circuit
	Ringer telephone		Oscilloscope	**VR**	Crystal (diode, breakdown)
E	Antenna	**MG**	Motor-Generator		Voltage regulator
	Insulator	**MK**	Microphone	**W**	Bus bar
	Magnet	**P**	Connector (plug)		Cable
	Post (binding)		Plug		Cable assembly
F	Fuse	**PS**	Power supply		Transmission
FL	Filter	**PU**	Head (recording)		
FS	Fire alarm		Head (playback)		
G	Generator		Pickup		

METRIC-INCH EQUIVALENTS

INCHES		MILLI-METERS	INCHES		MILLI-METERS
FRACTIONS	DECIMALS		FRACTIONS	DECIMALS	
	.00394	.1	15/32	.46875	11.9063
	.00787	.2		.47244	12.00
	.01181	.3	31/64	.484375	12.3031
1/64	.015625	.3969	1/2	.5000	12.70
	.01575	.4		.51181	13.00
	.01969	.5	33/64	.515625	13.0969
	.02362	.6	17/32	.53125	13.4938
	.02756	.7	35/64	.546875	13.8907
1/32	.03125	.7938		.55118	14.00
	.0315	.8	9/16	.5625	14.2875
	.03543	.9	37/64	.578125	14.6844
	.03937	1.00		.59055	15.00
3/64	.046875	1.1906	19/32	.59375	15.0813
1/16	.0625	1.5875	39/64	.609375	15.4782
5/64	.078125	1.9844	5/8	.625	15.875
	.07874	2.00		.62992	16.00
3/32	.09375	2.3813	41/64	.640625	16.2719
7/64	.109375	2.7781	21/32	.65625	16.6688
	.11811	3.00		.66929	17.00
1/8	.125	3.175	43/64	.671875	17.0657
9/64	.140625	3.5719	11/16	.6875	17.4625
5/32	.15625	3.9688	45/64	.703125	17.8594
	.15748	4.00		.70866	18.00
11/64	.171875	4.3656	23/32	.71875	18.2563
3/16	.1875	4.7625	47/64	.734375	18.6532
	.19685	5.00		.74803	19.00
13/64	.203125	5.1594	3/4	.7500	19.05
7/32	.21875	5.5563	49/64	.765625	19.4469
15/64	.234375	5.9531	25/32	.78125	19.8438
	.23622	6.00		.7874	20.00
1/4	.2500	6.35	51/64	.796875	20.2407
17/64	.265625	6.7469	13/16	.8125	20.6375
	.27559	7.00		.82677	21.00
9/32	.28125	7.1438	53/64	.828125	21.0344
19/64	.296875	7.5406	27/32	.84375	21.4313
5/16	.3125	7.9375	55/64	.859375	21.8282
	.31496	8.00		.86614	22.00
21/64	.328125	8.3344	7/8	.875	22.225
11/32	.34375	8.7313	57/64	.890625	22.6219
	.35433	9.00		.90551	23.00
23/64	.359375	9.1281	29/32	.90625	23.0188
3/8	.375	9.525	59/64	.921875	23.4157
25/64	.390625	9.9219	15/16	.9375	23.8125
	.3937	10.00		.94488	24.00
13/32	.40625	10.3188	61/64	.953125	24.2094
27/64	.421875	10.7156	31/32	.96875	24.6063
	.43307	11.00		.98425	25.00
7/16	.4375	11.1125	63/64	.984375	25.0032
29/64	.453125	11.5094	1	1.0000	25.4001

CONVERSIONS OF WEIGHTS & MEASURES

KNOWN	MULTIPLY BY	TO GET
CENTIMETERS (cm)	0.0328	FEET
CENTIMETERS	0.3937	INCHES
CENTIMETERS	0.01	METERS
CIRCLE	360°	DEGREES
CIRCULAR INCHES	0.7854	SQUARE INCHES
CUBIC FEET (cu ft)	1728.0	CUBIC INCHES
CUBIC FEET	0.0283	CUBIC METERS
CUBIC FEET	0.0370	CUBIC YARDS
CUBIC INCHES	0.00058	CUBIC FEET
CUBIC METERS (cu m)	35.3133	CUBIC FEET
DECIMETERS	3.937	INCHES
DECIMETERS	0.01	METERS
DECAMETERS	393.7	INCHES
DECAMETERS	10.0	METERS
DEGREES (deg. or °)	60.0	MINUTES
FAHRENHEIT	$\frac{5\,(°F - 32)}{9}$	CELSIUS, DEGREE
FEET (ft)	30.4801	CENTIMETERS
FEET	0.3048	METERS
INCHES (in)	2.5400	CENTIMETERS
INCHES	0.0254	METERS
INCHES	1000.0	MILS
KILOCYCLES	1000.0	CYCLES PER SECOND
KILOWATT-HOURS	1.3414	HORSEPOWER-HOURS
MEGACYCLES	1,000,000.0	CYCLES PER SECOND
METERS	3.2808	FEET
METERS	39.370	INCHES
MILLIMETERS (mm)	0.03937	INCHES
MILLIMETERS	0.001	METERS
MILLIMETERS	1000.0	MICRONS
MILLIMETERS	39.37	MILS
MILS	0.001	INCHES
MILS	25.4001	MICRONS
MILS	0.0254	MILLIMETERS
MINUTES (min)	60.0	SECONDS
SECONDS	0.01667	MINUTES
SQUARE FEET (sq ft)	144.0	SQUARE INCHES
SQUARE FEET	0.0929	SQUARE METERS
SQUARE FEET	0.1111	SQUARE YARDS
SQUARE INCHES (sq in)	0.00694	SQUARE FEET
SQUARE INCHES	645.1625	SQUARE MILLIMETERS
SQUARE INCHES	0.00077	SQUARE YARDS
SQUARE METERS (sq m)	10.7639	SQUARE FEET
SQUARE METERS	1.1960	SQUARE YARDS
SQUARE MILLIMETERS (sq mm)	0.00155	SQUARE INCHES
SQUARE MILLIMETERS	0.000001	SQUARE METERS

INCH TO METRIC
FRACTIONAL SIZES

in.		mm
1/16	=	1.587
1/8	=	3.175
3/16	=	4.762
1/4	=	6.350
5/16	=	7.937
3/8	=	9.525
7/16	=	11.112
1/2	=	12.700
9/16	=	14.287
5/8	=	15.875
11/16	=	17.462
3/4	=	19.050
13/16	=	20.637
7/8	=	22.225
15/16	=	23.812
1	=	25.400

RESISTOR COLOR CODE			
COLOR	1st DIGIT	2nd DIGIT	MULTIPLIER
BLACK	0	0	1
BROWN	1	1	10
RED	2	2	100
ORANGE	3	3	1,000
YELLOW	4	4	10,000
GREEN	5	5	100,000
BLUE	6	6	1,000,000
VIOLET	7	7	10,000,000
GRAY	8	8	100,000,000
WHITE	9	9	1,000,000,000
GOLD	-	-	.1
SILVER	-	-	.01

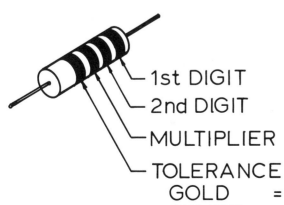

1st DIGIT
2nd DIGIT
MULTIPLIER
TOLERANCE
GOLD   = ± 5%
SILVER = ± 10%
NO BAND = ± 20%

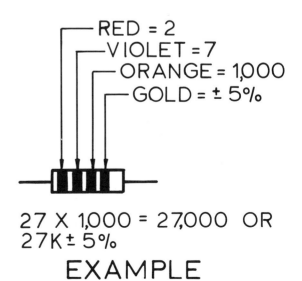

RED = 2
VIOLET = 7
ORANGE = 1,000
GOLD = ± 5%

27 X 1,000 = 27,000 OR
27K ± 5%

EXAMPLE

MILLIMETER-INCH CONVERSIONS

EXAMPLE:

CONVERT 2468.135 MILLIMETRES TO INCHES.

2000.	mm	=	78.74016 IN.
460.	mm	=	18.11024 IN.
8.13	mm	=	.32008 IN.
.005	mm	=	.00020 IN.
2468.135	mm	=	97.17068 IN.

INCREMENTS OF 1000 MILLIMETRES   1000 – 9000 MILLIMETRES

	1000	2000	3000	4000	5000	6000	7000	8000	9000
	39.37008	78.74016	118.11024	157.48031	196.85039	236.22047	275.59055	314.96063	354.33071

INCREMENTS OF 10 MILLIMETRES   0 – 1090 MILLIMETRES

	0	10	20	30	40	50	60	70	80	90
0	0	.39370	.78740	1.18110	1.57480	1.96850	2.36220	2.75591	3.14961	3.54331
100	3.93701	4.33071	4.72441	5.11811	5.51181	5.90551	6.29921	6.69291	7.08661	7.48031
200	7.87402	8.26772	8.66142	9.05512	9.44882	9.84252	10.23622	10.62992	11.02362	11.41732
300	11.81102	12.20472	12.59843	12.99213	13.38583	13.77953	14.17323	14.56693	14.96063	15.35433
400	15.74803	16.14173	16.53543	16.92913	17.32283	17.71654	18.11024	18.50394	18.89764	19.29134
500	19.68504	20.07874	20.47244	20.86614	21.25984	21.65354	22.04724	22.44094	22.83465	23.22835
600	23.62205	24.01575	24.40945	24.80315	25.19685	25.59055	25.98425	26.37795	26.77165	27.16535
700	27.55906	27.95276	28.34646	28.74016	29.13386	29.52756	29.92126	30.31496	30.70866	31.10236
800	31.49606	31.88976	32.28346	32.67717	33.07087	33.46457	33.85827	34.25197	34.64567	35.03937
900	35.43307	35.82677	36.22047	36.61417	37.00787	37.40157	37.79528	38.18898	38.58268	38.97638
1000	39.37008	39.76378	40.15748	40.55118	40.94488	41.33858	41.73228	42.12598	42.51969	42.91339

INCREMENTS OF .01 MILLIMETRES   0 – 10.09 MILLIMETRES

	.00	.01	.02	.03	.04	.05	.06	.07	.08	.09
0	0	.00039	.00079	.00118	.00157	.00197	.00236	.00276	.00315	.00354
.1	.00394	.00433	.00472	.00512	.00551	.00591	.00630	.00669	.00709	.00748
.2	.00787	.00827	.00866	.00906	.00945	.00984	.01024	.01063	.01102	.01142
.3	.01181	.01220	.01260	.01299	.01339	.01378	.01417	.01457	.01496	.01535
.4	.01575	.01614	.01654	.01693	.01732	.01772	.01811	.01850	.01890	.01929
.5	.01969	.02008	.02047	.02087	.02126	.02165	.02205	.02244	.02283	.02323
.6	.02362	.02402	.02441	.02480	.02520	.02559	.02598	.02638	.02677	.02717
.7	.02756	.02795	.02835	.02874	.02913	.02953	.02992	.03031	.03071	.03110
.8	.03150	.03189	.03228	.03268	.03307	.03346	.03386	.03425	.03465	.03504
.9	.03543	.03583	.03622	.03661	.03701	.03740	.03780	.03819	.03858	.03898
1.0	.03937	.03976	.04016	.04055	.04094	.04134	.04173	.04213	.04252	.04291
1.1	.04331	.04370	.04409	.04449	.04488	.04528	.04567	.04606	.04646	.04685
1.2	.04724	.04764	.04803	.04843	.04882	.04921	.04961	.05000	.05039	.05079
1.3	.05118	.05157	.05197	.05236	.05276	.05315	.05354	.05394	.05433	.05472
1.4	.05512	.05551	.05591	.05630	.05669	.05709	.05748	.05787	.05827	.05866
1.5	.05906	.05945	.05984	.06024	.06063	.06102	.06142	.06181	.06220	.06260
1.6	.06299	.06339	.06378	.06417	.06457	.06496	.06535	.06575	.06614	.06654
1.7	.06693	.06732	.06772	.06811	.06850	.06890	.06929	.06969	.07008	.07047
1.8	.07087	.07126	.07165	.07205	.07244	.07283	.07323	.07362	.07402	.07441
1.9	.07480	.07520	.07559	.07598	.07638	.07677	.07717	.07756	.07795	.07835
2.0	.07874	.07913	.07953	.07992	.08031	.08071	.08110	.08150	.08189	.08228
2.1	.08268	.08307	.08346	.08386	.08425	.08465	.08504	.08543	.08583	.08622
2.2	.08661	.08701	.08740	.08780	.08819	.08858	.08898	.08937	.08976	.09016
2.3	.09055	.09094	.09134	.09173	.09213	.09252	.09291	.09331	.09370	.09409
2.4	.09449	.09488	.09528	.09567	.09606	.09646	.09685	.09724	.09764	.09803
2.5	.09843	.09882	.09921	.09961	.10000	.10039	.10079	.10118	.10157	.10197
2.6	.10236	.10276	.10315	.10354	.10394	.10433	.10472	.10512	.10551	.10591
2.7	.10630	.10669	.10709	.10748	.10787	.10827	.10866	.10906	.10945	.10984
2.8	.11024	.11063	.11102	.11142	.11181	.11220	.11260	.11299	.11339	.11378
2.9	.11417	.11457	.11496	.11535	.11575	.11614	.11654	.11693	.11732	.11772
3.0	.11811	.11850	.11890	.11929	.11969	.12008	.12047	.12087	.12126	.12165
3.1	.12205	.12244	.12283	.12323	.12362	.12402	.12441	.12480	.12520	.12559
3.2	.12598	.12638	.12677	.12717	.12756	.12795	.12835	.12874	.12913	.12953
3.3	.12992	.13031	.13071	.13110	.13150	.13189	.13228	.13268	.13307	.13346
3.4	.13386	.13425	.13465	.13504	.13543	.13583	.13622	.13661	.13701	.13740
3.5	.13780	.13819	.13858	.13898	.13937	.13976	.14016	.14055	.14094	.14134
3.6	.14173	.14213	.14252	.14291	.14331	.14370	.14409	.14449	.14488	.14528
3.7	.14567	.14606	.14646	.14685	.14724	.14764	.14803	.14843	.14882	.14921
3.8	.14961	.15000	.15039	.15079	.15118	.15157	.15197	.15236	.15276	.15315
3.9	.15354	.15394	.15433	.15472	.15512	.15551	.15591	.15630	.15669	.15709
4.0	.15748	.15787	.15827	.15866	.15906	.15945	.15984	.16024	.16063	.16102
4.1	.16142	.16181	.16220	.16260	.16299	.16339	.16378	.16417	.16457	.16496
4.2	.16535	.16575	.16614	.16654	.16693	.16732	.16772	.16811	.16850	.16890
4.3	.16929	.16969	.17008	.17047	.17087	.17126	.17165	.17205	.17244	.17283
4.4	.17323	.17362	.17402	.17441	.17480	.17520	.17559	.17598	.17638	.17677
4.5	.17717	.17756	.17795	.17835	.17874	.17913	.17953	.17992	.18031	.18071
4.6	.18110	.18150	.18189	.18228	.18268	.18307	.18346	.18386	.18425	.18465
4.7	.18504	.18543	.18583	.18622	.18661	.18701	.18740	.18780	.18819	.18858
4.8	.18898	.18937	.18976	.19016	.19055	.19094	.19134	.19173	.19213	.19252
4.9	.19291	.19331	.19370	.19409	.19449	.19488	.19528	.19567	.19606	.19646
5.0	.19685	.19724	.19764	.19803	.19843	.19882	.19921	.19961	.20000	.20039
5.1	.20079	.20118	.20157	.20197	.20236	.20276	.20315	.20354	.20394	.20433
5.2	.20472	.20512	.20551	.20591	.20630	.20669	.20709	.20748	.20787	.20827
5.3	.20866	.20906	.20945	.20984	.21024	.21063	.21102	.21142	.21181	.21220
5.4	.21260	.21299	.21339	.21378	.21417	.21457	.21496	.21535	.21575	.21614
5.5	.21654	.21693	.21732	.21772	.21811	.21850	.21890	.21929	.21969	.22008
5.6	.22047	.22087	.22126	.22165	.22205	.22244	.22283	.22323	.22362	.22402
5.7	.22441	.22480	.22520	.22559	.22598	.22638	.22677	.22717	.22756	.22795
5.8	.22835	.22874	.22913	.22953	.22992	.23031	.23071	.23110	.23150	.23189
5.9	.23228	.23268	.23307	.23346	.23386	.23425	.23465	.23504	.23543	.23976
6.0	.23622	.23661	.23701	.23740	.23780	.23819	.23858	.23898	.23937	.23976
6.1	.24016	.24055	.24094	.24134	.24173	.24213	.24252	.24291	.24331	.24370
6.2	.24409	.24449	.24488	.24528	.24567	.24606	.24646	.24685	.24724	.24764
6.3	.24803	.24843	.24882	.24921	.24961	.25000	.25039	.25079	.25118	.25157
6.4	.25197	.25236	.25276	.25315	.25354	.25394	.25433	.25472	.25512	.25551
6.5	.25591	.25630	.25669	.25709	.25748	.25787	.25827	.25866	.25906	.25945
6.6	.25984	.26024	.26063	.26102	.26142	.26181	.26220	.26260	.26299	.26339
6.7	.26378	.26417	.26457	.26496	.26535	.26575	.26614	.26654	.26693	.26732
6.8	.26772	.26811	.26850	.26890	.26929	.26969	.27008	.27047	.27087	.27126
6.9	.27165	.27205	.27244	.27283	.27323	.27362	.27402	.27441	.27480	.27520
7.0	.27559	.27598	.27638	.27677	.27717	.27756	.27795	.27835	.27874	.27913
7.1	.27953	.27992	.28031	.28071	.28110	.28150	.28189	.28228	.28268	.28307
7.2	.28346	.28386	.28425	.28465	.28504	.28543	.28583	.28622	.28661	.28701
7.3	.28740	.28780	.28819	.28858	.28898	.28937	.28976	.29016	.29055	.29094
7.4	.29134	.29173	.29213	.29252	.29291	.29331	.29370	.29409	.29449	.29488
7.5	.29528	.29567	.29606	.29646	.29685	.29724	.29764	.29803	.29843	.29882
7.6	.29921	.29961	.30000	.30039	.30079	.30118	.30157	.30197	.30236	.30276
7.7	.30315	.30354	.30394	.30433	.30472	.30512	.30551	.30591	.30630	.30669
7.8	.30709	.30748	.30787	.30827	.30866	.30906	.30945	.30984	.31024	.31063
7.9	.31102	.31142	.31181	.31220	.31260	.31299	.31339	.31378	.31417	.31457
8.0	.31496	.31535	.31575	.31614	.31654	.31693	.31732	.31772	.31811	.31850
8.1	.31890	.31929	.31969	.32008	.32047	.32087	.32126	.32165	.32205	.32244
8.2	.32283	.32323	.32362	.32402	.32441	.32480	.32520	.32559	.32598	.32638
8.3	.32677	.32717	.32756	.32795	.32835	.32874	.32913	.32953	.32992	.33031
8.4	.33071	.33110	.33150	.33189	.33228	.33268	.33307	.33346	.33386	.33425
8.5	.33465	.33504	.33543	.33583	.33622	.33661	.33701	.33740	.33780	.33819
8.6	.33858	.33898	.33937	.33976	.34016	.34055	.34094	.34134	.34173	.34213
8.7	.34252	.34291	.34331	.34370	.34409	.34449	.34488	.34528	.34567	.34606
8.8	.34646	.34685	.34724	.34764	.34803	.34843	.34882	.34921	.34961	.35000
8.9	.35039	.35079	.35118	.35157	.35197	.35236	.35276	.35315	.35354	.35394
9.0	.35433	.35472	.35512	.35551	.35591	.35630	.35669	.35709	.35748	.35787
9.1	.35827	.35866	.35906	.35945	.35984	.36024	.36063	.36102	.36142	.36181
9.2	.36220	.36260	.36299	.36339	.36378	.36417	.36457	.36496	.36535	.36575
9.3	.36614	.36654	.36693	.36732	.36772	.36811	.36850	.36890	.36929	.36969
9.4	.37008	.37047	.37087	.37126	.37165	.37205	.37244	.37283	.37323	.37362
9.5	.37402	.37441	.37480	.37520	.37559	.37598	.37638	.37677	.37717	.37756
9.6	.37795	.37835	.37874	.37913	.37953	.37992	.38031	.38071	.38110	.38150
9.7	.38189	.38228	.38268	.38307	.38346	.38386	.38425	.38465	.38504	.38543
9.8	.38583	.38622	.38661	.38701	.38740	.38780	.38819	.38858	.38898	.38937
9.9	.38976	.39016	.39055	.39094	.39134	.39173	.39213	.39252	.39291	.39331
10.0	.39370	.39409	.39449	.39488	.39528	.39567	.39606	.39646	.39685	.39724

INCREMENTS OF .001 MILLIMETRES   .001 – .009 MILLIMETRES

	.001	.002	.003	.004	.005	.006	.007	.008	.009
	.00004	.00008	.00012	.00016	.00020	.00024	.00028	.00031	.00035

ROUND OFF PRACTICE:

THE TOTAL TOLERANCE APPLIED TO A MILLIMETRE DIMENSION SHALL BE THE BASIS FOR THE ACCURACY IN ROUNDING OFF DIMENSIONS AND TOLERANCES CONVERTED TO INCHES. TOTAL TOLERANCE VALUES AND REQUIRED ACCURACY FOR ROUNDING OFF ARE SHOWN IN THE FIGURES BELOW.

TOTAL TOLERANCE IN MILLIMETRES		CONVERTED VALUE IN INCHES SHALL BE ROUNDED TO
AT LEAST	LESS THAN	
0.0000	0.10	4 PLACES (.0001)
0.10	1.0	3 PLACES (.001)
1.0	--	2 PLACES (.01)

WHEN FIRST DIGIT DROPPED IS:	THE LAST DIGIT RETAINED IS:
LESS THAN 5	UNCHANGED
MORE THAN 5	INCREASED BY 1
5 FOLLOWED ONLY BY ZEROS	UNCHANGED IF EVEN
	INCREASED BY 1 IF ODD

EXAMPLE:

CONVERT 48.25 ±0.25 MILLIMETRES TO INCHES

CONVERT THE DIMENSION

40.	mm	=	1.57480 IN.
8.25	mm	=	.32480 IN.
48.25	mm	=	1.89960 IN.

CONVERT THE TOLERANCE

±0.25 mm = .00984 IN.

CONVERTED DIM & TOL

1.89960 ±.00984 IN.

ROUND OFF TO 3 PLACES BASED ON TOTAL TOLERANCE OF 0.50 mm

ROUNDED OFF DIM & TOL

1.900 ±.010 IN.

MICRO-METRE (MICRON)	MICRO-INCH	MICRO-METRE (MICRON)	MICRO-INCH
0.025	1	0.40	16
0.050	2	0.50	20
0.075	3	0.63	25
0.100	4	0.80	32
0.125	5	1.00	40
0.15	6	1.25	50
0.20	8	1.6	63
0.25	10	2.0	80
0.32	13	2.5	100

CONVERSION OF SURFACE TEXTURE DESIGNATIONS.

EXAMPLE:

1.6 / MICROMETRES = 63 / MICROINCHES

(Caterpillar Tractor Co.)

INCH-MILLIMETER CONVERSIONS

EXAMPLE:

CONVERT 135.7924 INCHES TO MILLIMETRES.

100.	IN.	=	2540.0000 mm
35.	IN.	=	889.0000 mm
.792	IN.	=	20.1168 mm
.0004	IN.	=	.01016 mm
135.7924	IN.	=	3449.12696 mm

ROUND OFF PRACTICE:

THE TOTAL TOLERANCE APPLIED TO AN INCH DIMENSION SHALL BE THE BASIS FOR THE ACCURACY IN ROUNDING OFF DIMENSIONS AND TOLERANCES CONVERTED TO MILLIMETRES. TOTAL TOLERANCE VALUES AND REQUIRED ACCURACY FOR ROUNDING OFF ARE SHOWN IN THE FIGURES BELOW.

TOTAL TOLERANCE IN INCHES		CONVERTED VALUE IN MILLIMETRES SHALL BE ROUNDED TO
AT LEAST	LESS THAN	
.0000	.0004	4 PLACES (.0001)
.0004	.004	3 PLACES (.001)
.004	.04	2 PLACES (.01)
.04	--	1 PLACES (.1)

WHEN FIRST DIGIT DROPPED IS:	THE LAST DIGIT RETAINED IS:
LESS THAN 5	UNCHANGED
MORE THAN 5	INCREASED BY 1
5 FOLLOWED ONLY BY ZEROS	UNCHANGED IF EVEN
	INCREASED BY 1 IF ODD

EXAMPLE:

CONVERT 3.655 ±.002 INCHES TO MILLIMETRES.

CONVERT THE DIMENSION

3.	IN.	= 76.2000 mm
.655	IN.	= 16.6370 mm
3.655	IN.	= 92.8370 mm

CONVERT THE TOLERANCE

±.002 IN. = ±.0508 mm

CONVERTED DIM & TOL

92.8370 ±.0508 mm

ROUND OFF TO 2 PLACES BASED ON TOTAL TOLERANCE OF .004 INCH.

ROUNDED OFF DIM & TOL

92.84 ±.05 mm

MICRO-INCH	MICROMETRE (MICRON)	MICRO-INCH	MICROMETRE (MICRON)
1	0.025	16	0.40
2	0.050	20	0.50
3	0.075	25	0.63
4	0.100	32	0.80
5	0.125	40	1.00
6	0.15	50	1.25
8	0.20	63	1.6
10	0.25	80	2.0
13	0.32	100	2.5

CONVERSION OF SURFACE TEXTURE DESIGNATIONS.

EXAMPLE:

63 ⎷ MICROINCHES = 1.6 ⎷ MICROMETRES

INCREMENTS OF 100 INCHES — 100 TO 900 INCHES

100	200	300	400	500	600	700	800	900
2540.0000	5080.0000	7620.0000	10160.0000	12700.0000	15240.0000	17780.0000	20320.0000	22860.0000

INCREMENTS OF 1 INCH — 1 TO 109 INCHES

	0	1	2	3	4	5	6	7	8	9
0	0	25.4000	50.8000	76.2000	101.6000	127.0000	152.4000	177.8000	203.2000	228.6000
10	254.0000	279.4000	304.8000	330.2000	355.6000	381.0000	406.4000	431.8000	457.2000	482.6000
20	508.0000	533.4000	558.8000	584.2000	609.6000	635.0000	660.4000	685.8000	711.2000	736.6000
30	762.0000	787.4000	812.8000	838.2000	863.6000	889.0000	914.4000	939.8000	965.2000	990.6000
40	1016.0000	1041.4000	1066.8000	1092.2000	1117.6000	1143.0000	1168.4000	1193.8000	1219.2000	1244.6000
50	1270.0000	1295.4000	1320.8000	1346.2000	1371.6000	1397.0000	1422.4000	1447.8000	1473.2000	1498.6000
60	1524.0000	1549.4000	1574.8000	1600.2000	1625.6000	1651.0000	1676.4000	1701.8000	1727.2000	1752.6000
70	1778.0000	1803.4000	1828.8000	1854.2000	1879.6000	1905.0000	1930.4000	1955.8000	1981.2000	2006.6000
80	2032.0000	2057.4000	2082.8000	2108.2000	2133.6000	2159.0000	2184.4000	2209.8000	2235.2000	2260.6000
90	2286.0000	2311.4000	2336.8000	2362.2000	2387.6000	2413.0000	2438.4000	2463.8000	2489.2000	2514.6000
100	2540.0000	2565.4000	2590.8000	2616.2000	2641.6000	2667.0000	2692.4000	2717.8000	2743.2000	2768.6000

INCREMENTS OF .001 INCH — .001 TO 1.009 INCHES

	.000	.001	.002	.003	.004	.005	.006	.007	.008	.009
0	0	.0254	.0508	.0762	.1016	.1270	.1524	.1778	.2032	.2286
.01	.2540	.2794	.3048	.3302	.3556	.3810	.4064	.4318	.4572	.4826
.02	.5080	.5334	.5588	.5842	.6096	.6350	.6604	.6858	.7112	.7366
.03	.7620	.7874	.8128	.8382	.8636	.8890	.9144	.9398	.9652	.9906
.04	1.0160	1.0414	1.0668	1.0922	1.1176	1.1430	1.1684	1.1938	1.2192	1.2446
.05	1.2700	1.2954	1.3208	1.3462	1.3716	1.3970	1.4224	1.4478	1.4732	1.4986
.06	1.5240	1.5494	1.5748	1.6002	1.6256	1.6510	1.6764	1.7018	1.7272	1.7526
.07	1.7780	1.8034	1.8288	1.8542	1.8796	1.9050	1.9304	1.9558	1.9812	2.0066
.08	2.0320	2.0574	2.0828	2.1082	2.1336	2.1590	2.1844	2.2098	2.2352	2.2606
.09	2.2860	2.3114	2.3368	2.3622	2.3876	2.4130	2.4384	2.4638	2.4892	2.5146
.10	2.5400	2.5654	2.5908	2.6162	2.6416	2.6670	2.6924	2.7178	2.7432	2.7686
.11	2.7940	2.8194	2.8448	2.8702	2.8956	2.9210	2.9464	2.9718	2.9972	3.0226
.12	3.0480	3.0734	3.0988	3.1242	3.1496	3.1750	3.2004	3.2258	3.2512	3.2766
.13	3.3020	3.3274	3.3528	3.3782	3.4036	3.4290	3.4544	3.4798	3.5052	3.5306
.14	3.5560	3.5814	3.6068	3.6322	3.6576	3.6830	3.7084	3.7338	3.7592	3.7846
.15	3.8100	3.8354	3.8608	3.8862	3.9116	3.9370	3.9624	3.9878	4.0132	4.0386
.16	4.0640	4.0894	4.1148	4.1402	4.1656	4.1910	4.2164	4.2418	4.2672	4.2926
.17	4.3180	4.3434	4.3688	4.3942	4.4196	4.4450	4.4704	4.4958	4.5212	4.5466
.18	4.5720	4.5974	4.6228	4.6482	4.6736	4.6990	4.7244	4.7498	4.7752	4.8006
.19	4.8260	4.8514	4.8768	4.9022	4.9276	4.9530	4.9784	5.0038	5.0292	5.0546
.20	5.0800	5.1054	5.1308	5.1562	5.1816	5.2070	5.2324	5.2578	5.2832	5.3086
.21	5.3340	5.3594	5.3848	5.4102	5.4356	5.4610	5.4864	5.5118	5.5372	5.5626
.22	5.5880	5.6134	5.6388	5.6642	5.6896	5.7150	5.7404	5.7658	5.7912	5.8166
.23	5.8420	5.8674	5.8928	5.9182	5.9436	5.9690	5.9944	6.0198	6.0452	6.0706
.24	6.0960	6.1214	6.1468	6.1722	6.1976	6.2230	6.2484	6.2738	6.2992	6.3246
.25	6.3500	6.3754	6.4008	6.4262	6.4516	6.4770	6.5024	6.5278	6.5532	6.5786
.26	6.6040	6.6294	6.6548	6.6802	6.7056	6.7310	6.7564	6.7818	6.8072	6.8326
.27	6.8580	6.8834	6.9088	6.9342	6.9596	6.9850	7.0104	7.0358	7.0612	7.0866
.28	7.1120	7.1374	7.1628	7.1882	7.2136	7.2390	7.2644	7.2898	7.3152	7.3406
.29	7.3660	7.3914	7.4168	7.4422	7.4676	7.4930	7.5184	7.5438	7.5692	7.5946
.30	7.6200	7.6454	7.6708	7.6962	7.7216	7.7470	7.7724	7.7978	7.8232	7.8486
.31	7.8740	7.8994	7.9248	7.9502	7.9756	8.0010	8.0264	8.0518	8.0772	8.1026
.32	8.1280	8.1534	8.1788	8.2042	8.2296	8.2550	8.2804	8.3058	8.3312	8.3566
.33	8.3820	8.4074	8.4328	8.4582	8.4836	8.5090	8.5344	8.5598	8.5852	8.6106
.34	8.6360	8.6614	8.6868	8.7122	8.7376	8.7630	8.7884	8.8138	8.8392	8.8646
.35	8.8900	8.9154	8.9408	8.9662	8.9916	9.0170	9.0424	9.0678	9.0932	9.1186
.36	9.1440	9.1694	9.1948	9.2202	9.2456	9.2710	9.2964	9.3218	9.3472	9.3726
.37	9.3980	9.4234	9.4488	9.4742	9.4996	9.5250	9.5504	9.5758	9.6012	9.6266
.38	9.6520	9.6774	9.7028	9.7282	9.7536	9.7790	9.8044	9.8298	9.8552	9.8806
.39	9.9060	9.9314	9.9568	9.9822	10.0076	10.0330	10.0584	10.0838	10.1092	10.1346
.40	10.1600	10.1854	10.2108	10.2362	10.2616	10.2870	10.3124	10.3378	10.3632	10.3886
.41	10.4140	10.4394	10.4648	10.4902	10.5156	10.5410	10.5664	10.5918	10.6172	10.6426
.42	10.6680	10.6934	10.7188	10.7442	10.7696	10.7950	10.8204	10.8458	10.8712	10.8966
.43	10.9220	10.9474	10.9728	10.9982	11.0236	11.0490	11.0744	11.0998	11.1252	11.1506
.44	11.1760	11.2014	11.2268	11.2522	11.2776	11.3030	11.3284	11.3538	11.3792	11.4046
.45	11.4300	11.4554	11.4808	11.5062	11.5316	11.5570	11.5824	11.6078	11.6332	11.6586
.46	11.6840	11.7094	11.7348	11.7602	11.7856	11.8110	11.8364	11.8618	11.8872	11.9126
.47	11.9380	11.9634	11.9888	12.0142	12.0396	12.0650	12.0904	12.1158	12.1412	12.1666
.48	12.1920	12.2174	12.2428	12.2682	12.2936	12.3190	12.3444	12.3698	12.3952	12.4206
.49	12.4460	12.4714	12.4968	12.5222	12.5476	12.5730	12.5984	12.6238	12.6492	12.6746
.50	12.7000	12.7254	12.7508	12.7762	12.8016	12.8270	12.8524	12.8778	12.9032	12.9286
.51	12.9540	12.9794	13.0048	13.0302	13.0556	13.0810	13.1064	13.1318	13.1572	13.1826
.52	13.2080	13.2334	13.2588	13.2842	13.3096	13.3350	13.3604	13.3858	13.4112	13.4366
.53	13.4620	13.4874	13.5128	13.5382	13.5636	13.5890	13.6144	13.6398	13.6652	13.6906
.54	13.7160	13.7414	13.7668	13.7922	13.8176	13.8430	13.8684	13.8938	13.9192	13.9446
.55	13.9700	13.9954	14.0208	14.0462	14.0716	14.0970	14.1224	14.1478	14.1732	14.1986
.56	14.2240	14.2494	14.2748	14.3002	14.3256	14.3510	14.3764	14.4018	14.4272	14.4526
.57	14.4780	14.5034	14.5288	14.5542	14.5796	14.6050	14.6304	14.6558	14.6812	14.7066
.58	14.7320	14.7574	14.7828	14.8082	14.8336	14.8590	14.8844	14.9098	14.9352	14.9606
.59	14.9860	15.0114	15.0368	15.0622	15.0876	15.1130	15.1384	15.1638	15.1892	15.2146
.60	15.2400	15.2654	15.2908	15.3162	15.3416	15.3670	15.3924	15.4178	15.4432	15.4686
.61	15.4940	15.5194	15.5448	15.5702	15.5956	15.6210	15.6464	15.6718	15.6972	15.7226
.62	15.7480	15.7734	15.7988	15.8242	15.8496	15.8750	15.9004	15.9258	15.9512	15.9766
.63	16.0020	16.0274	16.0528	16.0782	16.1036	16.1290	16.1544	16.1798	16.2052	16.2306
.64	16.2560	16.2814	16.3068	16.3322	16.3576	16.3830	16.4084	16.4338	16.4592	16.4846
.65	16.5100	16.5354	16.5608	16.5862	16.6116	16.6370	16.6624	16.6878	16.7132	16.7386
.66	16.7640	16.7894	16.8148	16.8402	16.8656	16.8910	16.9164	16.9418	16.9672	16.9926
.67	17.0180	17.0434	17.0688	17.0942	17.1196	17.1450	17.1704	17.1958	17.2212	17.2466
.68	17.2720	17.2974	17.3228	17.3482	17.3736	17.3990	17.4244	17.4498	17.4752	17.5006
.69	17.5260	17.5514	17.5768	17.6022	17.6276	17.6530	17.6784	17.7038	17.7292	17.7546
.70	17.7800	17.8054	17.8308	17.8562	17.8816	17.9070	17.9324	17.9578	17.9832	18.0086
.71	18.0340	18.0594	18.0848	18.1102	18.1356	18.1610	18.1864	18.2118	18.2372	18.2626
.72	18.2880	18.3134	18.3388	18.3642	18.3896	18.4150	18.4404	18.4658	18.4912	18.5166
.73	18.5420	18.5674	18.5928	18.6182	18.6436	18.6690	18.6944	18.7198	18.7452	18.7706
.74	18.7960	18.8214	18.8468	18.8722	18.8976	18.9230	18.9484	18.9738	18.9992	19.0246
.75	19.0500	19.0754	19.1008	19.1262	19.1516	19.1770	19.2024	19.2278	19.2532	19.2786
.76	19.3040	19.3294	19.3548	19.3802	19.4056	19.4310	19.4564	19.4818	19.5072	19.5326
.77	19.5580	19.5834	19.6088	19.6342	19.6596	19.6850	19.7104	19.7358	19.7612	19.7866
.78	19.8120	19.8374	19.8628	19.8882	19.9136	19.9390	19.9644	19.9898	20.0152	20.0406
.79	20.0660	20.0914	20.1168	20.1422	20.1676	20.1930	20.2184	20.2438	20.2692	20.2946
.80	20.3200	20.3454	20.3708	20.3962	20.4216	20.4470	20.4724	20.4978	20.5232	20.5486
.81	20.5740	20.5994	20.6248	20.6502	20.6756	20.7010	20.7264	20.7518	20.7772	20.8026
.82	20.8280	20.8534	20.8788	20.9042	20.9296	20.9550	20.9804	21.0058	21.0312	21.0566
.83	21.0820	21.1074	21.1328	21.1582	21.1836	21.2090	21.2344	21.2598	21.2852	21.3106
.84	21.3360	21.3614	21.3868	21.4122	21.4376	21.4630	21.4884	21.5138	21.5392	21.5646
.85	21.5900	21.6154	21.6408	21.6662	21.6916	21.7170	21.7424	21.7678	21.7932	21.8186
.86	21.8440	21.8694	21.8948	21.9202	21.9456	21.9710	21.9964	22.0218	22.0472	22.0726
.87	22.0980	22.1234	22.1488	22.1742	22.1996	22.2250	22.2504	22.2758	22.3012	22.3266
.88	22.3520	22.3774	22.4028	22.4282	22.4536	22.4790	22.5044	22.5298	22.5552	22.5806
.89	22.6060	22.6314	22.6568	22.6822	22.7076	22.7330	22.7584	22.7838	22.8092	22.8346
.90	22.8600	22.8854	22.9108	22.9362	22.9616	22.9870	23.0124	23.0378	23.0632	23.0886
.91	23.1140	23.1394	23.1648	23.1902	23.2156	23.2410	23.2664	23.2918	23.3172	23.3426
.92	23.3680	23.3934	23.4188	23.4442	23.4696	23.4950	23.5204	23.5458	23.5712	23.5966
.93	23.6220	23.6474	23.6728	23.6982	23.7236	23.7490	23.7744	23.7998	23.8252	23.8506
.94	23.8760	23.9014	23.9268	23.9522	23.9776	24.0030	24.0284	24.0538	24.0792	24.1046
.95	24.1300	24.1554	24.1808	24.2062	24.2316	24.2570	24.2824	24.3078	24.3332	24.3586
.96	24.3840	24.4094	24.4348	24.4602	24.4856	24.5110	24.5364	24.5618	24.5872	24.6126
.97	24.6380	24.6634	24.6888	24.7142	24.7396	24.7650	24.7904	24.8158	24.8412	24.8666
.98	24.8920	24.9174	24.9428	24.9682	24.9936	25.0190	25.0444	25.0698	25.0952	25.1206
.99	25.1460	25.1714	25.1968	25.2222	25.2476	25.2730	25.2984	25.3238	25.3492	25.3746
1.00	25.4000	25.4254	25.4508	25.4762	25.5016	25.5270	25.5524	25.5778	25.6032	25.6286

INCREMENTS OF .0001 INCH — .0001 TO .0009 INCH

.0001	.0002	.0003	.0004	.0005	.0006	.0007	.0008	.0009
.00254	.00508	.00762	.01016	.01270	.01524	.01778	.02032	.02286

(Caterpillar Tractor Co.)

INDEX

Z